Resilience of Critical Infrastructure Systems

Taylor and Francis Series in Resilience and Sustainability in Civil, Mechanical, Aerospace and Manufacturing Engineering Systems

Series Editor
Mohammad Noori
Cal Poly San Luis Obispo
Published Titles

Resilience of Critical Infrastructure Systems
Emerging Developments and Future Challenges
Zhishen Wu, Xilin Lu, and Mohammad Noori

For more information about this series, please visit: www.crcpress.com

Resilience of Critical Infrastructure Systems
Emerging Developments and Future Challenges

Edited by
Zhishen Wu, Xilin Lu, Mohammad Noori

CRC Press
Taylor & Francis Group
Boca Raton London New York

CRC Press is an imprint of the
Taylor & Francis Group, an **informa** business

CRC Press
Taylor & Francis Group
6000 Broken Sound Parkway NW, Suite 300
Boca Raton, FL 33487-2742

© 2020 by Taylor & Francis Group, LLC
CRC Press is an imprint of Taylor & Francis Group, an Informa business

No claim to original U.S. Government works

Printed on acid-free paper

International Standard Book Number-13: 978-0-367-47738-7 (Hardback)

This book contains information obtained from authentic and highly regarded sources. Reasonable efforts have been made to publish reliable data and information, but the author and publisher cannot assume responsibility for the validity of all materials or the consequences of their use. The authors and publishers have attempted to trace the copyright holders of all material reproduced in this publication and apologize to copyright holders if permission to publish in this form has not been obtained. If any copyright material has not been acknowledged, please write and let us know so we may rectify it in any future reprint.

Except as permitted under U.S. Copyright Law, no part of this book may be reprinted, reproduced, transmitted, or utilized in any form by any electronic, mechanical, or other means, now known or hereafter invented, including photocopying, microfilming, and recording, or in any information storage or retrieval system, without written permission from the publishers.

For permission to photocopy or use material electronically from this work, please access www.copyright.com (http://www.copyright.com/) or contact the Copyright Clearance Center, Inc. (CCC), 222 Rosewood Drive, Danvers, MA 01923, 978-750-8400. CCC is a not-for-profit organization that provides licenses and registration for a variety of users. For organizations that have been granted a photocopy license by the CCC, a separate system of payment has been arranged.

Trademark Notice: Product or corporate names may be trademarks or registered trademarks, and are used only for identification and explanation without intent to infringe.

Visit the Taylor & Francis Web site at
http://www.taylorandfrancis.com

and the CRC Press Web site at
http://www.crcpress.com

Contents

Preface vii
Contributors xi

Chapter 1 Introduction: Challenges and Generic Research Questions For Future Research On Resilience 1

Michel Bruneau, Gian-Paolo Cimellaro, Max Didier, Marco Domaneschi, Ivo Häring, Xilin Lu, Aftab Mufti, Mohammad Noori, Jinpin Ou, Anastasios Sextos, Shamim Sheikh, Ertugrul Taciroglu, Zhishen Wu, Lili Xie, Teruhiko Yoda, and Ying Zhou

Chapter 2 Resilience of Civil Infrastructure in a Life-Cycle Context 43

You Dong and Dan M. Frangopol

Chapter 3 Christchurch: Rebuilding a Resilient City? 49

Michel Bruneau and Gregory MacRae

Chapter 4 Resilient Bridge Decks Based on ISIS Winnipeg Principles 57

Aftab Mufti

Chapter 5 Resilience Considerations of a Historical Timber Bridge 67

Teruhiko Yoda and Weiwei Lin

Chapter 6 Resiliency and Recoverability of Concrete Structures 79

Zhishen Wu and Mohamed F.M. Fahmy

Chapter 7 Urban Infrastructures Resilience Assessing: An Overview & New Resilience Evaluation Theory 109

Wael A. Altabey, Mohammad Noori, and Ying Zhao

Chapter 8 Resilient Isolation-Structure Systems with Super-Large Displacement Friction Pendulum Bearings 123

Jinping Ou, Peisong Wu, and Xinchun Guan

Chapter 9 Real-Time City-Scale Time-History Analysis and Its Application in Resilience-Oriented Earthquake Emergency Response 141

Xinzheng Lu, Qingle Cheng, Zhen Xu, Yongjia Xu, and Chujin Sun

Chapter 10 Functionality Analyses of Engineering Systems: One Step toward Seismic Resilience 163

Tao Wang, Qingxue Shang, and Jichao Li

Chapter 11 Resilience of Bridges in Infrastructural Networks 177

Marco Domaneschi

Chapter 12 Building Resilience and Sustainability in Concrete Structures with FRP 189

Shamim A. Sheikh

Chapter 13 Resilience-Oriented Displacement-Based Seismic Design Procedure and Its Application in Self-Centering Metallic Energy-Dissipating Structures 203

Lu Liu and Bin Wu

Chapter 14 A Decision-Making Framework for Enhancing Resilience of Road Networks in Earthquake Regions 213

Anastasios Sextos and Ioannis Kilanitis

Index 225

Preface

This book entitled *Resilience of Critical Infrastructure Systems: Emerging Developments and Future Challenges* is a collection of selected papers presented at The 2nd International Workshop on Resilience held on October 31–November 1, 2018, in Nanjing, and on November 2, 2018, in Shanghai, China. This successful invitational workshop was jointly organized by Southeast University and Tongji University, and was sponsored by a number of professional societies and organizations.

The workshop brought together about 30 keynote and invited speakers and over 100 participants from civil, mechanical, systems, and earthquake engineering, security systems, and risk and reliability assessment to brainstorm and to develop more effective approaches for the grand challenges facing the resilience of critical infrastructure systems, rapidly emerging and growing filed. The keynote and invited presentations emphasized the important role that the advances in fundamental and applied research in the fields that will play a major role in the future research of the resilience of critical infrastructure. In particular, research in the areas of smart technologies and smart cities, which are complementary to the research into the resilience of critical infrastructure.

Some of the key recommendations of the workshop included:

- It is critically important that we develop new strategies and processes to assure that increasingly interconnected critical infrastructure systems become more resilient and withstand the threats of man-made and mega natural hazards, and extreme events.
- The ever increasing pace of urban development requires we build more resilient critical infrastructure and communities by considering uncertainty quantification, advanced statistical techniques, and probabilistic methods.
- The emerging concept of smart cities, and the research achievements in structural health monitoring provide an opportunity to combine advances in modeling and smart technologies. This will provide an opportunity for groundbreaking discoveries to improve the resilience of critical infrastructure systems and to transform infrastructure from physical structures to responsive systems.
- It is important to incorporate societal issues, and community resilience in a comprehensive and strategic plan for tackling the impacts of natural and man-made hazards, and to make sound research investments to better understand the interaction of critical infrastructure and socio-economic factors. In this regard, it is important to utilize data sciences, data analytics, and artificial intelligence to facilitate a multi-disciplinary collaboration.

This two-day workshop included both keynote and invited lectures to promote an awareness of the state-of-the-art in the field of critical infrastructure resilience and to identify the challenges that lie ahead. To facilitate the identification of critical issues to be addressed, three working groups were formed. Working groups' discussions identified pressing grand challenges facing the resilience of critical infrastructure that require interdisciplinary research roadmaps that aim to aggressively resolve those grand challenges. Following thematic areas were collectively identified for the brainstorming sessions:

- Working Group 1 – Fundamental concepts and framework for resilient critical infrastructure.
- Working Group 2 – Intelligent and innovative technologies to enhance structural resiliency.
- Working Group 3 – How to incorporate a multi-disciplinary approach that considers societal, community, and human factors in critical infrastructure resilience research by employing data science and artificial intelligence tools.

The working group discussions were strongly shaped by the cross-disciplinary research and educational perspectives of the workshop participants. The grand challenge problems identified should be considered as the priority issues facing critical infrastructure research in the coming decades. The workshop recommendations focused on greater adoption of multi-disciplinary research strategies, embracing mega natural and man-made hazards, socio-economic, intelligent technologies, and data analytics.

It is envisioned that the implementation of the suggested research roadmaps proposed by these working groups will result in safer and more resilient urban communities, and stronger economies, all of which are factors in enhancing the overall quality of life of our global society. Equally important is the development of new integrated interdisciplinary programs for the education of the next generation of engineers so they are well qualified to undertake the proposed research roadmap. Finally, implementation of the solutions will only be successful if all stakeholders in academia, industry, and government work collaboratively in nurturing, implementing, and maintaining the solutions.

The participants unanimously supported the following resolution:

The International Workshop on Resilience should be established as a standing workshop to be held every two years. This ensures the workshop can serve as the primary catalyst for continuous assessment of the research directions in the resilience of critical infrastructure systems. This workshop can also become a conduit to promote the global significance of this emerging field and for aggressively pursuing the pressing grand challenges.

The workshop participants unanimously approved a proposal for the next workshop to be held in 2020. The International Society for Health Monitoring of Intelligent Infrastructure (ISHMII) played an active role in the success of this workshop and will continue to play a major role in the future plans for establishing this workshop as a standing international forum for the exchange and development of research in the multi-disciplinary field of critical infrastructure. Thus, it was concluded that ISHMII should make the commitment to continue to be the sponsor of the future series of this workshop and to reach out to other organizations to seek their involvement, support, and participation.

Preface

Workshop Co-Chairs and the Organizing Committee:

Zhishen Wu, Southeast University (Co-Chair)
Xilin Lu, Tongji University (Co-Chair)
Mohammad Noori, Cal Poly, San Luis Obispo (Secretary General)
Jian Zhang, Southeast University
Ying Zhou, Tongji University
Gian Paolo Cimellaro, Politecnico di Torino
Ertugrul Taciroglu, University of California, Los Angles
Jianguo Cai, Southeast University

Guest Editor – Professor Zhishen Wu is a Professor at Southeast University, China, and Ibaraki University, Japan. His research expertise includes structural/concrete/maintenance engineering and advanced composite materials. He is the author or co-author of over 600 refereed papers including over 200 journal articles and has given 50 keynote or invited papers. He also holds 50 patents. Dr. Wu was awarded the JSCE Research Prize by Japan Society of Civil Engineering in 1990, the JSCM Technology Award by the Japan Society for Composite Materials in 2005 and 2009, SHM person of the year Award by the International Journal of Structural Health Monitoring, and the National Prize for Progress in Science and Technology (2nd) of China in 2012. He is also a member of the Japan National Academy of Engineering. He chairs the China Chemical Fibers Association Committee on Basalt Fibers, and is the President of the International Society for Structural Health Monitoring of Intelligent Infrastructure (ISHMII). He is also an elected fellow of ASCE, JSCE, ISHMII, and the International Institute of FRP in Construction. Moreover, he serves as an editor, associate editor, and editorial board member for more than ten international journals, including the founding chief editor of the *International Journal of Sustainable Materials and Structural Systems*. Professor Wu has supervised a large number of national research projects in China and was the founder of the International of Institute for Urban Systems Engineering, at Southeast University.

Guest Editor – Prof. Xilin Lu is currently a Distinguished Professor of Civil Engineering at Tongji University. He conducted extensive research during 1990 at the University of Alberta, Canada, on the punching shear behavior of RC slabs, and worked as a Research Associate at the University of Hong Kong from 1991 to 1992, where he investigated the seismic behavior of reinforced concrete shear walls with vertical slits. His research areas include dynamic modeling, the testing and detailed analysis of complex structures, and seismic design and response control of tall buildings. Professor Lu's work has resulted in the development of seismic design codes of the Shanghai Municipal Government and China National Code. He is the chief editor of the *Structural Design of Tall and Special Buildings Journal* published by Wiley; he is the Vice President of the International Association for Experimental Structural Engineering, a Fellow of the International Association for Bridge and Structural Engineering, and a Member of ASCE. He has published over 100 journal papers, presented more than 30 keynote or invited lectures at international conferences, and published over 200 papers in international conference proceedings. He received the

2017 Nathan M. Newmark Medal "for pioneering research in the development of dynamic testing, structural seismic design, and response control for tall buildings," as well as "for his commitment to education and professional service."

Corresponding Editor – Professor Mohammad Noori is a Professor of Mechanical Engineering at Cal Poly, San Luis Obispo, a Fellow of the American Society of Mechanical Engineering, and a recipient of the Japan Society for Promotion of Science Fellowship. Noori's work in nonlinear random vibrations, seismic isolation, and the application of artificial intelligence methods for structural health monitoring is widely cited. He has authored over 250 refereed papers, including over 100 journal articles, 6 scientific books, and he has edited 25 technical, and special journal, volumes. Noori has supervised over 90 graduate students and post-doc scholars, and has presented over 100 keynote, plenary, and invited talks. He is the founding executive editor of international journal of sustainable materials and structural systems (ISMSS) and has served on the editorial board of over ten other journals and as a member of numerous scientific and advisory boards. He has been a distinguished visiting professor at several highly ranked global universities, and directed the Sensors Program at the National Science Foundation in 2014. He has been a founding director or co-founder of three industry-university research centers and held chair professorships at two major universities and has served a member of the advisory boards of the college of engineering at several highly ranked universities. He served as the Dean of Engineering at Cal Poly for five years, as the Chair of the national committee of mechanical engineering department heads, and was one of seven co-founders of the National Institute of Aerospace, in partnership with NASA Langley Research Center. Noori is an elected member of Sigma Xi, Pi Tau Sigma, Chi-Epsilon, and Sigma Mu Epsilon engineering honorary societies.

Contributors

Wael A. Altabey
Nanjing Zhixing Information
 Technology Co., Ltd.
Nanjing, China,
Department of Mechanical Engineering,
 Faculty of Engineering, Alexandria
 University, Alexandria, Egypt
and
Department of Mechanical Engineering
Faculty of Engineering
Alexandria University
Alexandria, Egypt

Michel Bruneau
University at Buffalo
Buffalo, New York

Qingle Cheng
Beijing Engineering Research Center
 of Steel and Concrete Composite
 Structures
Tsinghua University
Beijing, China

Gian-Paolo Cimellaro
Department of Structural, Geotechnical,
 and Building Engineering
Polytechnic of Turin
Turin, Italy

Max Didier
Department of Civil, Geomatic
 and Environmental Engineering
 (D-BAUG)
ETH Zurich
Zurich, Switzerland

Marco Domaneschi
Polytechnic of Turin
Department of Structural, Geotechnical,
 and Building Engineering
Turin, Italy

You Dong
Department of Civil and Environmental
 Engineering
The Hong Kong Polytechnic
 University
Kowloon, Hong Kong

Mohamed F.M. Fahmy
International Institute for Urban
 Systems Engineering
Southeast University
Nanjing, China
and
Civil Engineering Department
Assiut University
Assiut, Egypt

Dan M. Frangopol
Department of Civil and Environmental
 Engineering
Engineering Research Center for
 Advanced Technology for Large
 Structural Systems (ATLSS Center)
Lehigh University
Bethlehem, Pennsylvania

Xinchun Guan
Key Lab of Structures Dynamic
 Behavior and Control
Ministry of Education
Harbin Institute of Technology
Harbin, China

Ivo Häring
Department of Civil, Geomatic
 and Environmental Engineering
 (D-BAUG)
ETH Zurich
Zurich, Switzerland

Ioannis Kilanitis
Aristotle University of Thessaloniki
Thessaloniki, Greece

Jichao Li
Key Laboratory of Earthquake
  Engineering and Engineering
  Vibration
Institute of Engineering Mechanics
China Earthquake Administration
Harbin, China

Weiwei Lin
Department of Civil Engineering
Aalto University School of Engineering
Espoo, Finland

Lu Liu
Harbin Institute of Technology at
  Weihai
Weihai, China

Xilin Lu
Research Institute of Structural
  Engineering and Disaster Reduction
Tongji University
Shanghai, China

Xinzheng Lu
Key Laboratory of Civil Engineering
  Safety and Durability
China Education Ministry
Department of Civil Engineering
Tsinghua University
Beijing, China.

Gregory MacRae
University of Canterbury
Christchurch, New Zealand

Aftab Mufti
University of Manitoba
Winnipeg, Manitoba, Canada

Mohammad Noori
Department of Mechanical Engineering
  and ASME Fellow
California Polytechnic State University
San Luis Obispo, California

Jinping Ou
Key Lab of Structures Dynamic
  Behavior and Control
Ministry of Education
Harbin Institute of Technology
Harbin, China

Anastasios Sextos
Department of Civil Engineering
University of Bristol, UK

Qingxue Shang
Key Laboratory of Earthquake
  Engineering and Engineering
  Vibration
Institute of Engineering Mechanics
China Earthquake Administration
Harbin, China

Shamim A. Sheikh
Department of Civil Engineering
University of Toronto
Toronto, Ontario, Canada

Chujin Sun
Beijing Engineering Research Center
  of Steel and Concrete Composite
  Structures
Tsinghua University
Beijing, China

Ertugrul Taciroglu
Department of Civil and Environmental
  Engineering
University of California, Los Angeles
Los Angeles, California

Tao Wang
Key Laboratory of Earthquake
  Engineering and Engineering
  Vibration
Institute of Engineering Mechanics
China Earthquake Administration
Harbin, China

Contributors

Bin Wu
Wuhan University of Technology
Wuhan, China

Peisong Wu
Key Lab of Structures Dynamic
  Behavior and Control
Ministry of Education
Harbin Institute of Technology
Harbin, China

Zhishen Wu
National and Local Unified Engineering
  Research Center for Basalt Fiber
  Production and Application
  Technology
International Institute for Urban
  Systems Engineering
Southeast University
Nanjing, China

Lili Xie
Department of Civil and Environmental
  Engineering
Waseda University
Tokyo, Japan

Yongjia Xu
Beijing Engineering Research Center
  of Steel and Concrete Composite
  Structures
Tsinghua University
Beijing, China

Zhen Xu
School of Civil and Environmental
  Engineering
University of Science and Technology
  Beijing
Beijing, China

Teruhiko Yoda
Department of Civil Engineering
Waseda University
Tokyo, Japan

Ying Zhao
International Institute for Urban
  Systems Engineering
Southeast University,
Nanjing, China

Ying Zhou
Department of Disaster Mitigation for
  Structures
College of Civil Engineering
Tongji University, China,

1 Introduction
Challenges and Generic Research Questions For Future Research On Resilience

*Michel Bruneau, Gian-Paolo Cimellaro,
Max Didier, Marco Domaneschi, Ivo Häring,
Xilin Lu, Aftab Mufti, Mohammad Noori,
Jinpin Ou, Anastasios Sextos, Shamim Sheikh,
Ertugrul Taciroglu, Zhishen Wu, Lili Xie,
Teruhiko Yoda, and Ying Zhou*

CONTENTS

1.1 Resilience of Critical Infrastructure 2
    1.1.1 Grand Challenges toward Future Resilient Socio Cyber-Technical Self-Learning Infrastructure Systems 3
    1.1.2 Contextualizing, Motivating, and Elaborating on Tasks for Road-Mapping of Future Resilience Research 6
    1.1.3 Operationalize Sample Questions for Road-Mapping Future Resilience Research 8
Literature 10
1.2 Frameworks, Fundamentals, and Education for Future Infrastructure Risk Control and Resilience 11
    1.2.1 Background and Introduction 12
    1.2.2 Objective Details 12
    1.2.3 Key Challenges for Fundamental Resilience Research and Education 13
    1.2.4 Major Research Gaps 14
    1.2.5 Framework to Address the Challenges 15
    1.2.6 Concepts, Methods, and Technologies to Be Further Developed 16
    1.2.7 Roadmaps and Strategies Proposed for Future Implementation 17
    1.2.8 Summary and Conclusions 17
Literature 18

1.3 How to Improve Critical Infrastructure Systems with Emerging
    Technologies: Predictive Simulation and Emerging Technologies 19
    1.3.1 Background and Introduction 20
    1.3.2 Critical Infrastructure Definitions and Main Objectives List 21
        1.3.2.1 TI – Transportation Infrastructure 22
        1.3.2.2 EI – Energy Infrastructure 22
        1.3.2.3 WW – Waste Water System 22
        1.3.2.4 ES – Emergency Services 22
        1.3.2.5 IT – Information Technologies 22
    1.3.3 BR – Building Structures and Residences Resisting
          Extreme Loads 22
    1.3.4 Key Challenges 23
    1.3.5 Better Understanding Intra and Interdependencies of Critical
          Infrastructure 24
    1.3.6 Data-Informed and Driven Infrastructure Modeling
          and Simulation 25
    1.3.7 Future Health Monitoring and Early Warning 25
    1.3.8 Resilience-Based Improvements for Critical Infrastructural
          Systems Already in Service 27
    1.3.9 Resilience-Based Design for Infrastructure and Systems,
          Community Resilience, and Demonstration Flagship Projects
          (Living Labs) 27
    1.3.10 Emerging Technologies for Innovation of Resilient
           Infrastructure and Systems in Design and Planning 28
    1.3.11 Roadmaps and Strategies Proposed for Future Implementation 29
    1.3.12 Summary 30
Literature 31
1.4 Big Data, Machine Learning and AI for Future Human Support in
    Disruption Events and Critical Infrastructure System Resilience 31
    1.4.1 Motivation, Background and Introduction 32
    1.4.2 Objectives 34
    1.4.3 Key Challenges 35
    1.4.4 Major Gaps in the State of the Art 36
    1.4.5 Framework to Address the Challenges 37
    1.4.6 Proposed Actions to Tackle the Challenges: Concepts, Methods,
          and Technologies for Future Resilience Research 38
    1.4.7 Roadmaps and Strategies Proposed for Future Implementation 39
    1.4.8 Summary and Conclusions 39
References 40

1.1 RESILIENCE OF CRITICAL INFRASTRUCTURE

At the second international workshop on resilience, IRW 2018, held in Nanjing and Shanghai, future research questions for critical infrastructure were discussed and exemplarily identified in three working groups. The present introductory chapter highlights some of the main drivers and challenges for future research on risk

Introduction

control of critical infrastructure systems (Section 1.1.1), relates to selected similar road-mapping approaches for motivating the overarching main questions posed to the working groups (Section 1.1.2), and attempts to formulate some more operational sample questions for each workshop topic (Section 1.1.3). Drivers of future resilience research for critical infrastructure systems include new, more, and multidimensional threats as well as more complex but also capable infrastructure systems, which are more blurry, extended, distributed, sensing, and intelligent and self-learning. Similar road-mapping approaches identified include top-level policy pieces, academic institutionalized consulting reports, a report on a workshop series on resilience engineering, consolidated scientific workshop results with similar ambitions, and inaugural journal founding articles that motivate further civil security research from a mainly technical perspective. The many listed key workshop questions and additional operational workshop questions are expected to be of value also for future critical infrastructure research road-mapping.

1.1.1 GRAND CHALLENGES TOWARD FUTURE RESILIENT SOCIO CYBER-TECHNICAL SELF-LEARNING INFRASTRUCTURE SYSTEMS

Resilience, originally discovered as the stunning property that allows individuals to better survive even the worst conditions in early youth, by now has made its way through many disciplines as a stimulating concept. Disciplines include psychology, ecological systems, science, community, and social sciences, as well as disciplines that have socio-technical systems as a subject of investigation but also purely technical systems of various scales. In particular, engineering and technical sciences that investigate critical infrastructure systems and related subsystems, and many more disciplines, are taking up the concept of resilience for innovating research. However, a concise technical and natural science-driven approach to resilience of the socio-technical that takes up sufficient perspectives of the humanities is still missing.

Resilience is often defined as the context-sensitive, individually, societally, juridically, etc. acceptable capability of systems to better prepare for, respond to, protect from, detect, prevent, respond, recover, and learn before, during, and after disruptive events. Correspondingly, resilience cycle phases can be defined, e.g.,

- Preparation
- Building up protective capabilities (e.g., of structures)
- Detection
- Prevention
- Absorption
- Response
- Recovery
- Adaption and Learning

Resilience is understood ultimately as a systemic property referring to a system functionality (system service) behavior of interest. However, for instance, it is not yet known how to determine more systematically which resilience cycle phases are relevant for given infrastructure and scenarios as well as how to derive main and secondary (infrastructure) system functions.

The above resilience definition often allows for efficient deductive approaches to system understanding and modeling, i.e., starting from expected system performance functions and their time-dependent behavior in cases of disruption. Up to now, it is not yet known how to take full advantage of the deductive analytical options of the resilience concept as compared to more bottom-up constructive approaches that leverage system modeling and system predictive simulative capabilities more as a byproduct of advanced modeling and simulation capabilities.

Major damaging events may occur suddenly or gradually and be man-made, anthropogenic (natural-technical, natech), criminal, sabotage, foreign-state, and agency driven, malicious, even terroristic events. Often multiple events need to be considered ordered in an attack vector. Most damaging disruptions of the more recent past were combined loading events. However, the anticipation of such events still seems much too optimistic, even when considering the bitter lessons learned.

Feasible combinations almost not yet considered include, for instance, long-term events (e.g., years-long drafts, heavy long-term rain, social unrest, and instability) combined with unfavorable short term conditions (e.g., hot days and winds, heavy winds, snow melting, heavy rain). Such events could be further engraved with malicious man-made attacks (e.g., arson, cyber-attacks, terror attacks on dams, attacks on supply logistics). More holistic approaches to mitigate these natural and man-made hazards are needed. Moreover, respective related resilience assessment schemes for multi-dimensional hazard mitigation strategies should be developed.

Acceptance refers to the overall behavior in all-resilience cycle phases but typically also includes the individual acceptance of each phase and for all types of potential disruptions. However, resilience acceptance is today rather a byproduct of specific applications and even only implicitly agreed upon rather than part of systematic resilience determination and improvement or development schemes, when compared, for instance, with risk control and management approaches. In the latter case, a cannon of quantities are available for risk visualization, comparison, and risk evaluation, including but not limited to individual local risk, individual risk profiles, local collective risks, and collective or group risks often displayed in F-N diagrams. Similar benchmark quantities that are well accepted by users are still missing in the resilience domain.

Often properties, capabilities, or technical capabilities are named that systems are expected to fulfill for being resilient such as sensing or detection capability, modeling and reality representation capability, inference making, acting as well as adaption, and learning. It remains an open question whether such rather technical intuitive capabilities are sufficiently specific to ensure the resilience of socio-technical systems, even if they strongly shift the discourse to system capabilities that can actively and generically be constructed rather than adding disruption-driven system improvements to ensure the resilience of systems.

Further, it is often emphasized that resilient systems need to be considered at all systems layers such as physical, technical, cyber, organizational, and societal-economical as well as environmental. The advantage of such sorting concepts is challenged by the systemic nature of the resilience concepts asking for the very avoidance of silo approaches. At the same time, the insight that successful systemic layers, e.g., physical robustness of buildings or cyber infrastructure, often don't take into account any challenging dependency discourses emphasizes the need to better understand system models and their implications.

Also, such properties as robustness, responsiveness, redundancy, resourcefulness, and rapidity ("the Rs") that can be linked to selected response-curve behaviors of interest are emphasized to be necessary for resilient systems. However, which system properties, which sufficiently good "Rs" are needed in given scenarios and which of them will lead to sufficient resilient systems is increasingly seen not to be generally answerable but to depend strongly on the system and decision making context, the system performance functions of interest, the related boundary conditions, and the specific properties of the system itself.

While the concept of resilience is increasingly taken up, tailored, and operationalized in various disciplines, the society as a whole and researchers are confronted with an ever-increasing complexity of systems. Systems are increasingly interconnected, smart, and system boundaries tend to become unclear. However, the question arises whether the many paradigm shifts these system character changes imply are taken up sufficiently, e.g., by leveraging more complex system modeling and simulation approaches.

Methods of machine learning (ML) and artificial intelligence (AI) render possible novel capabilities of systems while challenging the successful concept of deterministic system entities and interactions so important for classical system analysis, which is also at the core of many system resilience analysis approaches. Here new approaches are asked for to assess and take advantage of ML and AI approaches.

More complex and much less deterministic self-learning systems add to the continuously growing potential disruption types and the possibility that systems learn or even perfect their dangerous behavior, up the possibility of malicious intelligent systems. Compared with the very challenging latter threat of a system becoming malicious against humans, societal assets, or the environment, the threat that machine learning capabilities are systematically used to induce harm to systems by humans is a more recent but already well-established threat, nevertheless challenging.

In summary, as technological capabilities of technical solutions increase, the number of potential threats and hazards also increases. However, at the same time, it can be expected that especially sensing and smart system technologies, digital analytics, machine learning approaches up to artificial intelligence capabilities allow novel approaches for better understanding, quantifying, assessing, and evaluating the resilience of systems. In a similar way, such technologies can be expected to generate more resilient systems, including being self-resilient in the sense that additional machine learning approaches are used to avoid any unsafe behavior. Hence, the question also arises as to how much future resilient systems will need to rely on and leverage big data and analytics, machine learning, and AI capabilities and technologies.

Along these lines, in a more general way, the question arises whether increasing technological capabilities will also have the ability to control the increasing threat and disruption potentials of critical infrastructure systems.

1.1.2 Contextualizing, Motivating, and Elaborating on Tasks for Road-Mapping of Future Resilience Research

Taking advantage of the momentum of the 2nd International Resilience Workshop (IRW) 2018 at Nanjing and Shanghai, China, October 31 to November 2, 2018 [1], all the participants present at the workshop committed to contributing to outline future resilience research goals from a mainly engineering-technical perspective.

The participants inter alia had backgrounds in the fields of earthquake engineering, civil engineering of bridges and infrastructure, catastrophe management, network modeling, electrical engineering, quantitative risk analysis, technical and functional safety, and computer science. A wide range of methodologies was covered from conceptual frameworks, engineering-analytical approaches, component and large-scale structure resilience and their simulation, modeling, and simulation to network modeling sciences, machine learning, and artificial intelligence expert systems.

Taking this background into account, and for stimulating the formulation of detailed future resilience research roadmaps based on identified challenges, technological drives, and societal needs, the following guiding question complexes were identified before the workshop started [2]:

(1) How to enhance understanding of the fundamental processes underlying natural hazards and extreme events on various spatial and temporal scales, as well as the variability inherent in such hazards and events.
(2) How to improve our capability to model and forecast (including uncertainty quantifications of) such hazards and events by better understanding infrastructure, combined with advances in modeling and smart technologies that promise an opportunity for groundbreaking discoveries to improve resilience.
(3) How to enhance societal preparedness and societal resilience against the impacts of natural and man-made hazards by making sound research investments to better understand the technology that supports critical infrastructure and human–technology interactions.

The research questions can be related to the "shared research needs, breakthroughs, and observations about emerging threats and challenges to the security and resilience of the [United States of America] Nation's critical infrastructure" as identified at the National Science and Technology Council (NSTC) 2018 Critical Infrastructure Security and Resilience (CISR) Stakeholder Workshop, held on February 28, 2018, in Washington, DC. "Federal, state, private, and academic stakeholders" listed the following items [3]:

- Computational models, sensor networks, big data, and self-healing systems are promising approaches to solving CISR challenges.
- Techniques and technologies developed for one sector or system may be successfully adapted into others.
- Research and development (R&D) into the dynamics of interdependent systems and into human and social factors in critical infrastructure systems are of particular interest.

Introduction 7

In this context, the following road-mapping endeavors are also of interest:

(i) The US presidential directive statements on the resilience of critical infrastructures [4].
(ii) The German Acatech initiative ResilienTech [5].
(iii) The working group results in modeling and simulation approaches for critical infrastructure protection of a North Atlantic Treaty Organization (NATO) science workshop on critical infrastructure protection and resilience [6].
(iv) Another example is the "Foresight review of resilience engineering: Designing for the expected and unexpected" stipulated by Lloyd's Register Foundation, which provides background, definitions, and challenges while focusing on engineering solutions and covers a wide range of fields [7].
(v) Fundamental concepts, frameworks, resilience extreme value problems, resilience event propagation models, process models and methods, and tools for understanding all phases of the resilience cycle from preparation, protection, detection, prevention, loading, absorption, response, recovery, improvement (bouncing back better), and learning as well as taking account of all-resilience dimensions (see, e.g., [8], [9]).

During the first day of the workshop, the following amendments were added to the formulation of the tasks of the three working groups taking up the feedback of the participants [10]:

(1) **Framework and fundamentals of future infrastructure risk control and resilience:** How to enhance understanding of the fundamental processes underlying natural hazards, extreme events on various spatial and temporal scales, as well as the variability inherent in such hazards and events? What curricular changes are necessary to better prepare the future generation of civil engineers for critical infrastructure (CIS) research?

(2) **Critical infrastructure systems predictive simulation and emerging technologies for improvement:** How to improve our capability to model and forecast (including uncertainty quantifications of) such hazards and events with a better understanding of infrastructure socio cyber-physical systems? How to advance modeling and smart technologies that promise opportunities for groundbreaking discoveries to improve resilience? For instance, how to transform infrastructure, from physical structures to sensing, self-aware, and responsive systems? How to assure that increasingly interconnected CIS meet demands and withstand environmental hazards?

(3) **Leveraging big data, analytics, machine learning, and AI for human support and CIS resilience:** How to enhance societal preparedness and societal resilience against the impacts of natural and man-made hazards? How to make sound research investments to better develop technology that supports critical infrastructure and human–technology interactions? How to leverage in this context big data, AI platforms, and data analytics at

various scales? How to promote a multidisciplinary collaboration between the engineering, computer and information science, and social, behavioral, and economic sciences, also to address socio-political and technical issues?

1.1.3 Operationalize Sample Questions for Road-Mapping Future Resilience Research

Within the working group discussions, and during the evaluation of the workshop, it became apparent that the mostly orthogonal coverage but also complementary scope of each working group could be supported and further clarified with the following list of generic questions. These questions are also believed to provide potential guidance for critical infrastructure roadmaps beyond the horizon of the present workshop and of the immediate future.

(1.1) How to better understand and support the decision making context of infrastructure users, operators, and decision makers.

(1.2) How to better understand the selection of short-term and long-term risk control and resilience improvement measures?

(1.3) How to better understand the drivers for the selection of decision makers for insurance, state support, and management on exception regimes, and how they are separated in real-world contexts?

(1.4) What are fundamental ideas, concepts, procedural models, theories, models, approaches, and methods to better define, understand, represent, model, simulate, predict, improve, and optimize critical infrastructure systems regarding risk control and resilience?

(1.5) What are the resilience dimension, scales, levels of complexity, resolutions, and higher-order concepts to be considered?

(1.6) What are key quantities and metrics as well as methods, techniques, and measures and their interplay to improve the assessment, quantification, design, and development of sustainable risk control and resilience of technical systems, infrastructure (sub) systems, critical infrastructures as well as networks of such systems?

(1.7) How to develop sustainable solutions countering disruptions also in case of major damage effects?

(1.8) What are the hazard and threat types, the loading characteristics of natural, anthropogenic, natech, accidental, sabotage, intended, and terroristic loadings for which risks should be controlled and resilience be improved?

(1.9) What are the implications of these fundamental needs for future resilience engineering education from a mainly technical perspective?

(1.10) How can such future resilience research and education ensure sufficient interchange and input from the humanities?

(2.1) What are metadata of critical infrastructure models, boundary conditions, realistic resolutions, and coverage?

(2.2) How can relevant input data be gathered or generated using existing and novel sensors?

Introduction

(2.3) How can the data be made more easily accessible, taking advantage of existing sources?

(2.4) Which modeling approaches are suited, flexible, and sufficient for critical infrastructure systems modeling and predictive simulation?

(2.5) How can the coupling between infrastructure systems be covered in a predictive fashion?

(2.6) How to cope with bad, limited, truncated, reduced data sets?

(2.7) How to handle fundamental, deep, systematic, and statistic uncertainties of modeling approaches and data used?

(2.8) How to identify, model, and successfully design self-organizing systems' risk control and resilience of sets of independent subsystems without master–slave relations, in particular in Internet of Things (IoT) architectures, ad hoc networks, decentral system designs, and in case of emergent system behavior?

(2.9) How to address from a modeling perspective the limited access to information and data on infrastructure design, operation, and functional parameters as well as known weaknesses and flaws, in particular when going beyond single operator realms.

(2.10) Predictive and fast modeling of single infrastructure systems for the simulation of numerous events, in particular, for the application of advanced Monte Carlo approaches.

(2.11) How to further develop the layer of resilience approach for event propagation through multiple assessment layers, in particular for root cause and possible effects identification over multiple domains and levels of abstraction, through the development of unit expansion vectors (insertion of unity) for the identification of all possible event propagation trajectories.

(3.1) How to understand and model socio-technical interaction from a technical perspective in the context of critical infrastructure.

(3.2) How to take advantage of non-dedicated secondary sources of information regarding infrastructure health and status pre and post disruptive events, e.g., from citizens' smartphone sensors, pictures, and traffic flow data.

(3.3) How to better ensure a joint perspective on risk control and resilience as well as reliability, availability, maintenance, control, and optimization of technical systems, in particular how to avoid different approaches and implemented systems for each domain.

(3.4) How better determine the local loading of infrastructure systems, components, and elements for modeling on a high level of resolution (order of few meters and below) as a crucial input for better prediction and real-time risk assessment and prediction, data-driven as well as theoretical.

(3.5) How to leverage new digital infrastructure urban formats (e.g., CityGML) and build models (e.g., building information models (BIM)) for fast damage assessment and prediction.

(3.6) How to extract basic data and time-dependent health data of infrastructures and operators in all-resilience cycle phases from distributed, disparate, and heterogeneous data sources.

In addition, it was proposed to use the following report structure for each working group report [11]:

1. A brief background and introduction describing the motivations, including section structuring overview
2. Objectives list
3. Statement of key challenge items
4. A brief overview of major gaps of the state of the art
5. A framework of proposed options to tackle the challenges
6. Concepts, methods, and technologies to be further developed in the future
7. Tentative roadmaps and strategies for future implementation
8. Conclusions and summary

Within the three working groups the three topics are covered using the context provided in the present overview, guided by the set of questions to be addressed and the report structure proposed, see [12] [13] [14]. In addition, further resilience research questions and research challenges are listed.

LITERATURE

1. *2nd International Workshop on Resilience (IRW) 2018*, http://www.workshop-china 2018.resiltronics.org/, last access on December 27, 2018.
2. Workshop draft program as of October 8, 2018.
3. Subcommittee on Critical Infrastructure Security and Resilience, Committee on Homeland and National Security of the National Science and Technology Council. (February 28, 2018). *Summary of the 2018 Critical Infrastructure Security and Resilience Stakeholder Workshop.* https://www.whitehouse.gov/wp-content/uploads/2018/03/New-CISR-Stakeholder-Workshop-Summary-Formatted-FINAL.pdf, last access on December 2, 2018.
4. *National Security Strategy of the United States of America*, December 2017, The White House, Washington DC, https://www.whitehouse.gov/wp-content/uploads/2017/12/NSS-Final-12-18-2017-0905.pdf, last access on November 22, 2018.
5. Acatech – German National Academy of Science and Engineering. Resilience-by-design: a strategy for the technology issues of the future, acatech POSITION PAPER – Executive summary and recommendations. https://www.acatech.de/wp-content/uploads/2018/03/acatech_POSITION_RT_KF_eng_140508.pdf, last access on November 22, 2018.
6. I. Häring, G. Sansavini, E. Bellini, N. Martyn, T. Kovalenko, M. Kitsak, G. Vogelbacher, K. Ross, U. Bergerhausen, K. Barker, I. Linkov. (2017). Towards a generic resilience management, quantification and development process: general definitions, requirements, methods, techniques and measures, and case studies. In: *Resilience and Risk: Methods and Application in Environment, Cyber and Social Domains*, Editors: I. Linkov, J. M. Palma-Oliveira, pp. 21–80, Springer, ISBN 9789402411225, http://www.springer.com/de/book/9789402411225.
7. Lloyd's Register Foundation Report Series: No. 2015.2, Foresight review of resilience engineering: designing for the expected and unexpected, October 2015, http://www.lrfoundation.org.uk/publications/resilience-engineering.aspx, last access on November 25, 2018.
8. K. Thoma, B. Scharte, D. Hiller, T. Leismann. (2016). Resilience engineering as part of security research: definitions, concepts and science approaches. *Eur J Secur Res*, Volume 1, Issue 1, pp. 3–19. doi:10.1007/s41125-016-0002-4.

Introduction

9. I. Häring, S. Ebenhöch, A. Stolz. (2016). Quantifying resilience for resilience engineering of socio technical systems. *Eur J Secur Res*, Volume 1, Issue 1, pp. 21–58. doi:10.1007/s41125-015-0001-x.
10. Workshop questions as transferred to and agreed by the working groups on October 30, 2018.
11. Final workshop questions as shared between the working groups on October 30, 2018.
12. Lili Xie, Gian-Paolo Cimellaro, Michel Bruneau, Zhishen Wu, Max Didier, Mohammad Noori, Ivo Häring. (2018). Frameworks, fundamentals and education for future infrastructure risk control and resilience: workshop 1 report. *Proceedings of 2nd International Workshop on Resilience (IRW) 2018*, Nanjing and Shanghai, China.
13. Aftab Mufti, Xilin Lu, Jinpin Ou, Shamim Sheikh, Ying Zhou, Marco Domaneschi, Mohammad Noori, Ivo Häring. (2018). How to improve critical infrastructure systems with emerging technologies: future critical infrastructure systems predictive simulation and emerging technologies – Report of working group 2. *Proceedings of 2nd International Workshop on Resilience (IRW) 2018*, Nanjing and Shanghai, China.
14. Teruhiko Yoda, Ertugrul Taciroglu, Ivo Häring, Anastasios Sextos, Mohammad Noori. (2018). Big data analytics, machine learning and AI for future human support against disruption events and for critical infrastructure system resilience: report of working group 3. *Proceedings of 2nd International Workshop on Resilience (IRW) 2018*, Nanjing and Shanghai, China.

1.2 FRAMEWORKS, FUNDAMENTALS, AND EDUCATION FOR FUTURE INFRASTRUCTURE RISK CONTROL AND RESILIENCE

This section covers the identification of future fundamental resilience research needs and related academic educational challenges, as found at the 2nd International Workshop on Resilience 2018. Fundamental research needs were agreed to consist of the development of flexible and generally accepted frameworks, i.e., assessment process models, resilience improvement, development, implementation, and optimization models. Such models should also take the cultural, societal context, and expectations of operators, users, and citizens into account, in particular, how to translate them into acceptance and evaluation criteria. Much more work is believed to be necessary to understand and simulate local loadings (e.g., on building level in case of earthquakes and flooding), especially of combined multiple threats, such as physical impact, flood loading, or cyber attacks combined with physical and natural hazards. Measuring and metrics for resilience are believed to remain ongoing future tasks, in particular, on a system level. Specific challenges identified comprise megacities, legacy infrastructure, fast simulation of large-scale urban built environments as well as increasingly interlinked infrastructure systems, and the use of (unspecific and dedicated) data, data analytics up to self-learning approaches (machine learning, KI). Modern semantic digital building and infrastructure formats are expected to enable novel developments regarding (semi) automatic assessments. The education focus for long-term scientific and applied capacity improvement is believed to build on strong (multiple) MINT subject domain experts, who should also take advanced courses in system science modeling approaches as well as data-driven sciences. In all cases, students are proposed to be taught within broad real-world application projects to learn how to involve users and decision makers and respective participatory science approaches for enhancing future resilience research. Along with the main argumentative lines, this section provides lists of key questions and challenges and identifies similarly most prominent approaches and methods. It outlines a tentative timeline for the implementation of the future fundamental and education resilience research roadmap.

1.2.1 BACKGROUND AND INTRODUCTION

As threat modalities, variability, and demand levels to infrastructure systems of today's worldwide societies and communities are increasing, the need for advanced overall and sustainable risk control proves to be a fundamental prerequisite for thriving societal, individual, economic, and environmental well-being and development.

At the 2nd International Resilience Workshop (IRW) 2018 held at Nanjing and Shanghai [1], the experts present at the workshop committed to outlining future resilience research goals from a mainly engineering-technical perspective. To arrive at sufficient specific recommendations, as discussed in more detail in the introduction to the three workshop reports [2], the focus of working group 1 was to roadmap research needs regarding the enhanced understanding of fundamental processes underlying natural and extreme events on various spatial and temporal scales. This comprised of taking account of the variability inherent in such hazards and events, and on curricular changes necessary to better prepare the future generation of civil engineers for critical infrastructure resilience (CIS) research. Working Group 2 addresses advanced predictive modeling for critical infrastructure and related technology innovations, in particular, data gathering [3]. Workshop 3 covered the leverage of data analytics, machine learning up to AI approaches for better infrastructures, and the support of humans in case of disruption events [4].

Fundamental processes to be addressed within the research needs road-mapping were specified to cover advanced risk control and resilience management processes, risk, and resilience analysis, mechanical, technical, cyber, etc. properties, and behaviors of infrastructures at risk, as well as organizational issues, societal agendas, and even worldwide contexts framing the understanding of modern socio cyber-physical infrastructure systems including users, operators, and decision makers, from a (mainly technical and engineering) science and fundamental perspective, respectively.

The following report first defines fundamental infrastructure resilience research and education top-level goals (Section 1.2.2), from which key challenges (Section 1.2.3) are derived by contrasting with a research landscape review of the state of the art (Section 1.2.4). To ensure comprehensiveness, Section 1.2.5 provides a boundary condition aware framework for future fundamental and educational infrastructure resilience research needs. This sets the stage for concepts, methods, and technologies to be advanced in future resilience research (Section 1.2.6). Section 1.2.7 provides a tentative timing schedule for the identified research endeavors.

1.2.2 OBJECTIVE DETAILS

Working group 1 identified the following objectives regarding fundamental frameworks, research needs, and methodological gaps for future resilient infrastructures:

Introduction

- As acceptable overall risk control and resilience strongly depends on the societal context and consensus, it needs to be better clarified which overall frameworks, process models, stepwise-iterative approaches, and framing methods are necessary and sufficient for the
  - Contextualization,
  - Organizational and technical assessment,
  - Risk and resilience evaluation, decision making, and
  - Design and/or improvement of critical infrastructure systems.
- Better understanding of threats, their combinations (threat vectors), parametrization, and loading description, on high-resolution scales, especially combined classical, malicious, cyber, and intelligent threats.
- Better modeling and simulation of socio-cyber-physical infrastructure systems at risk, exposed to threats, and tested with respect to their resilience, taking account of inter and intra dependencies, and the blurry boundaries of systems.
- Provision of risk control and resilience quantities accepted by end-users and academia.
- Improve risk and resilience evaluation criteria and consensus, sufficient for processing engineering-technical risk and resilience quantities.
- Provision of fundamental principles, approaches, methods, and (structural and dynamic) solutions for better risk control and resilience, including fast but predictive models.
- Definition of curricula guidelines for future resilience research taking account of the high variability of subject domains and the need for specific knowledge to allow for progress.
- Addressing the need for continuous academic education.

1.2.3 Key Challenges for Fundamental Resilience Research and Education

To illustrate the overall objective of future-proof resilience research ambitions, the following specific challenges were identified, in part also in addition to the questions posed in Section 1.2.3:

- Resilience frameworks and processes need to address in a systematic way boundary contexts, e.g., decision making competences and resources.
- Especially technical science-driven resilience frameworks need to better take account of and extend resilience quantification beyond civil engineering or purely technical aspects to encapsulate social and economic impacts and dimensions, considering entire communities holistically.
- It needs to be made more explicit how types of country, cultures, communities, and local contexts determine the boundary conditions of resilience assessments, e.g., what is considered as resilient or safe versus not resilient/unsafe in different cultures.
- The advantages and disadvantages of the different functions of infrastructure versus system structural risk control and resilience approaches (i.e.,

whether to follow a more functional, or system dynamic versus a more static, or system structural, design approach) need to be better understood.
- There is a lack of risk control and resilience, overall chance and options qualitative, semi-quantitative, quantitative measures and quantities, metrics and aggregated evaluation options, in particular, such quantities that cover several scales of resolution from local components (structural members), to structures, buildings, quarters urban regions, countries, and worldwide regions.
- It needs to be identified at which scale resilience should be assessed, in particular, whether most relevant resolution levels are at the level of individual decision makers (e.g., house owners), at the community level (e.g., quarter major) or at higher political or social levels.
- Scaling and normalization of resilience metrics are open questions, e.g., how to compare regions with high hazard levels with low hazard regions.
- There is a lack of systematic and generally accepted approaches of resilience management, e.g., based on well-defined risks of not being resilient, are missing, in particular, how, if at all, to embed them in current risk assessment schemes.
- The specific challenges of megacities are not yet understood; for instance, spatial scales need to be covered and resolved. Positive as well as negative scaling effects are expected, e.g., when comparing the resilience with respect to local threats with the resilience with respect to non-localized threats, e.g., local terror events versus large-scale weather calamities. In the former, rescue forces will not be limited at all, whereas in the latter accumulation effects need to be considered.
- Account for legacy infrastructures, anticipate long-term use, allow for the replacement of sensitive aging (digital) parts of infrastructures.

So far, research on critical infrastructures in a civil context has not yet been canonized by any means. It is observed that true progress in this domain often depends on a broad knowledge in several disciplines. Typically advancing the domain requires the application of fundamental knowledge from related disciplines, e.g., civil engineering, mechanical engineering, computer science, physics, or other MINT subjects. Balancing the subject-specific fundamentals and the emerging interdisciplinary science is considered to be a challenging endeavor. It becomes even more challenging when considering the employability of developed models or acquired knowledge in non-academic domains.

1.2.4 Major Research Gaps

Resilience frameworks are often not ambitious enough regarding their scope and generalizability, but are, on the contrary, too generic to be applicable to different domains while keeping all their advantages. This occurs in particular if such frameworks and processes are de facto adapted to very specific applications, e.g., earthquake engineering or explosive terroristic threats only.

Current frameworks do not systematically aim at the maximum separation possible between resilience assessment and improvement process steps (see, e.g., the discussion in [5] within the initial sections), e.g., consider the use of the term scenario. On the other hand, schemes tend to miss interdependencies, e.g., consider the often much too implicit selection of resilience assessment quantities. For instance, it is not distinguished between management objectives, quantities that are computed for assessment, reference and comparison quantities, and resilience evaluation and acceptance steps.

The advantage of deductive and inverse methods in resilience event propagation root cause reduction and possible event identification has not yet been taken into account. Also, a systematic resilience dimensional analysis and a reduction to sufficient and necessary resilience dimensions are missing. How to relate risk and resilience management, especially in operational contexts, is still an often-discussed issue. Possible starting points and concepts of the questions discussed in this text section are given, for example, in the respective sections of [6].

Existing frameworks are typically tailored to well-known hazards, in particular for earthquake engineering. Examples include the MCEER framework of the Multidisciplinary Center for Earthquake Engineering Research of the University at Buffalo [7] and to a lesser extent the PEOPLES [8] framework. For instance, the schematic step-by-step procedure of the MCEER methodology proposes the following sequence: (1) Define extreme event scenarios (e.g., probabilistic seismic hazard analysis (PSHA), ground motion selection); (2) Define the system model; (3) Evaluate the response of the model; (4) Compute different performance measures (e.g., losses, recovery time, functionality, resilience); (5) Identify remedial mitigation actions (e.g., advanced technologies) and/or resilience actions (e.g., resourcefulness, redundancy, etc.); (6) Redesign the system. This approach strongly focuses on the engineering steps and does not provide explicit support of where to place context, evaluation, and decision making considerations.

Current damage models at various scales, from structures to buildings and infrastructures, are still mainly focusing on initial damage effects, rather than on recovery and even less post-event improvement options. It is expected that advanced modeling options will allow to better foresee how to rebuild faster or even rebuild better.

Current approaches too often fail to explain the application context, non-technical constraints, and the expected impact of the research results on improving risk control, resilience, and the decision making process.

Several Universities are already setting up or already offering curricula in the domain of critical infrastructure protection and resilience research and engineering. Examples include: (i) University at Buffalo, MCEER, a national multidisciplinary and multi-hazard earthquake engineering research center [9]; (ii) University of Freiburg, Department of Sustainable Systems Engineering (INATECH) with its thematic topic resilience engineering [10] focusing on transfer of research to industrial innovation [11].

1.2.5 Framework to Address the Challenges

Frameworks of improved risk control and resilience for critical infrastructure systems will need to better take into account the following topics:

- Existing standardizations and their gaps
- Seamless assessment of standard and non-standard operations

- Include societal and individual expectations more explicitly, as well as decision options and available resources
- Much more explicitness regarding overall risk acceptance criteria and societal priorities
- Allowance for participative and informed decision making of individuals, decision makers, and the representatives of the society
- Frameworks should take advantage of the available access to digitalized spatial data and semantic infrastructure data
- Frameworks should be modular and sufficiently specific to predict effects on the level of individual buildings
- Modeling and simulation approaches should be scalable, i.e., cover a broad range of modeling options from empirical-statistical, via engineering-analytical to simulative approaches in order to be able to handle the variability of the amount and quality of available data
- Modeling and simulation approaches should deliver their output adapted to the intended use of the results, e.g., detailed structural data versus traffic-light-assessments for volunteer rescue forces
- The framework should take account of social media data and computational resources, of increasing interconnectedness of devices and of local computing options
- Segregation and diversification of communication channels need to be taken into account
- Frameworks should address the level of education and experience required to conduct the necessary assessments and decisions

1.2.6 Concepts, Methods, and Technologies to Be Further Developed

Concepts, methods, and technologies expected to be most relevant and to be further developed in the future to improve risk control and resilience for critical infrastructure systems include:

- Large-scale simulations at the city and regional level
- Parallel computing, e.g., use of simplified access to graphic process units (GPUs) parallel computations via the CUDA API for fast real-time simulations [12]
- Use of spatial distribution of ground motion at the regional level and in spatially distributed infrastructures (water, power, gas distribution networks, etc.) with a high local resolution
- Soil-structure interaction effects at the regional scale and city scale. How different foundations of skyscrapers might interfere with the soil and modify the dynamic response
- Use and automatic analysis of digital pictures/images to collect model input data, to support the selection of models, to identify the damage, or the propagation of hazards, such as fire
- Use of Bayesian networks to model critical infrastructure system's service demand and recovery

Introduction

- Development of dynamic Bayesian probabilistic networks (BPNs) or agent-based models (ABMs) to model the recovery processes, for instance, sequences of such processes, repair times, while considering resource constraints
- Use of Weighted Bayesian updating, e.g., to quantify building stock fragility
- Use and analysis of big data as obtained, e.g., from twitter, smart grids, sensors in buildings, for a broad spectrum of tasks including the detection of threats, damaging events, health monitoring (pre- and post-event), anomaly detection, etc.
- Leverage of machine learning and artificial intelligence algorithms, e.g., random forests or neural networks to assess repair costs or to predict continuous degradation. Anticipated levels include
  - Informed application
  - Tailoring
  - Further development of methods

1.2.7 Roadmaps and Strategies Proposed for Future Implementation

It is expected that mainly existing frameworks and approaches will be further extended, interlinked, and enriched with new technological approaches and methods. Furthermore, standardization is expected to increase on different levels.

At the same time, dominating approaches are expected (platform effect, "the winner takes all") as standardization and digitalization of infrastructures increase as well as the interchange between formats becoming more and more automatized for structured data and more and more practicable even for unstructured data.

Especially such international standards as CityGML [13] on the semantic digital city, infrastructure, and building levels and building information models [14] on a single building level are candidate formats, not only for the exchange of digital data, but also for assessment procedures using such data. Another example are extended GIS formats as well as, for example, OpenStreetMap or similar proprietary formats.

The expectations of the prediction accuracy of models, along with associated uncertainty prediction, will further rise. At the same time, the computation time will be further increased, increasing the demand for advanced computing methods. As advanced computation requires more consideration and more event trajectories, computation resources are expected to remain a strong limiting factor for future decades.

Regarding time scales, reliable data gathering is expected to remain a limiting factor as well, due to large legacy infrastructure system fractions.

1.2.8 Summary and Conclusions

In summary, working group 1 was challenged to identify the future fundamentals of technical science-driven resilience research for improving critical infrastructure systems.

- It was identified that comprehensive frameworks and processes need to be further developed. To obtain stable technical assessments, quantifications, and solutions that adequately take up the true needs of all actors without anticipating solutions from a purely technical point of view are required.

Several approaches were identified as being of main interest for future resilient and resource-effective infrastructure solutions:

- Seamless reliability, failure, and disruption handling capability.
- Large-area simulation capability for a range of natural and man-made catastrophes, development with high resolution on local scales, e.g., for local building loading, taking into account known geophysics.
- Leverage of disciplinary methods for quantitative and probabilistic resilience research requires fundamental disciplinary research guided by true user needs.
- Need for the development of fast computation capabilities for multiple scenario analysis, using advanced computational approaches that also can be employed in case of emergencies.
- Advanced computational and/or engineering approaches include technical solutions, hierarchical models, abstract tailored models, and real-time modeling based on data.
- Overall life-cycle considerations including possible major disruptions as well as aging.
- The development of curricula for resilience engineering needs to be rooted in dedicated domain-specific anchor subjects, as well as generic capabilities such as complex systems modeling, graphical models, all methods of (classical) systems analysis and engineering.

LITERATURE

1. *2nd International Resilience Workshop (IRW) 2018*, Nanjing and Shanghai, China, October 30–November 2, 2018, http://www.workshop-china2018.resiltronics.org/, last access on November 23, 2018.
2. Lili Xie, Gian-Paolo Cimellaro, Michel Bruneau, Zhishen Wu, Max Didier, Mohammad Noori, Aftab Mufti, Xilin Lu, Jinpin Ou, Shamim Sheikh, Ying Zhou, Marco Domaneschi, Teruhiko Yoda, Ertugrul Taciroglu, Ivo Häring, Anastasios Sextos. (2018). Challenges and generic research questions for future research on resilience of critical infrastructure: introduction to workshop reports of the 2nd International Workshop on Resilience (IRW) 2018. *Proceedings of the 2nd International Resilience Workshop (IRW) 2018*, Nanjing and Shanghai, China.
3. Aftab Mufti, Xilin Lu, Jinpin Ou, Shamim Sheikh, Ying Zhou, Marco Domaneschi, Mohammad Noori, Ivo Häring. (2018). How to improve critical infrastructure systems with emerging technologies: future critical infrastructure systems predictive simulation and emerging technologies – Report of working group 2. *Proceedings of 2nd International Workshop on Resilience (IRW) 2018*, Nanjing and Shanghai, China.
4. Teruhiko Yoda, Ertugrul Taciroglu, Ivo Häring, Anastasios Sextos, Mohammad Noori. (2018). Big data analytics, machine learning and AI for future human support against disruption events and for critical infrastructure system resilience: report of working group 3. *Proceedings of 2nd International Workshop on Resilience (IRW) 2018*, Nanjing and Shanghai, China.

5. I. Häring, G. Sansavini, E. Bellini, N. Martyn, T. Kovalenko, M. Kitsak, G. Vogelbacher, K. Ross, U. Bergerhausen, K. Barker, I. Linkov. (2017). Towards a generic resilience management, quantification and development process: general definitions, requirements, methods, techniques and measures, and case studies. In: *Resilience and Risk: Methods and Application in Environment, Cyber and Social Domains*, Editors: I. Linkov, J. M. Palma-Oliveira, pp. 21–80, Springer, ISBN 9789402411225, http://www.springer.com/de/book/9789402411225.
6. Ivo Häring, Stefan Ebenhöch, Alexander Stolz. (2016). Quantifying resilience for resilience engineering of socio technical systems. *Eur J Secur Res*, Volume 1, Issue 1, pp. 21–58. doi: 10.1007/s41125-015-0001-x.
7. MCEER, University at Buffalo, http://www.buffalo.edu/mceer/about.html, last access on November 25, 2018.
8. Chris S. Renscher, Amy E. Frazier, Lucy A. Arendt, Gian-Paolo Cimellaro, Andrei M. Reinhorn, Michel Bruneau. (October 8, 2010). *A framework for defining and measuring resilience at the community scale: the PEOPLES resilience framework*. Technical Report MCEER-10-0006. http://mceer.buffalo.edu/publications/catalog/reports/A-Framework-for-Defining-and-Measuring-Resilience-at-the-Community-Scale-The-PEOPLES-Resilience-Framework-MCEER-10-0006.html, last access on November 25, 2018.
9. MCEER, http://www.buffalo.edu/mceer/about.html, last access on November 27, 2018.
10. INATEC, Department for Sustainable Systems Engineering, University of Freiburg, https://www.inatech.uni-freiburg.de/en?set_language=en, last access on November 27, 2018.
11. Sustainability Center Freiburg, http://www.leistungszentrum-nachhaltigkeit.de/en/sustainability-center/about-us, last access on November 27, 2018.
12. CUDA, https://developer.nvidia.com/cuda-zone, last access on November 25, 2018.
13. CityGML, https://www.citygml.org/, last access on November 25, 2018.
14. Nawari O. Nawari. (2018). *Building Information Modeling: Automated Code Checking and Compliance Processes*, CRC Press, ISBN 1498785336, https://www.crcpress.com/search/results?kw=ISBN+1498785336.

1.3 HOW TO IMPROVE CRITICAL INFRASTRUCTURE SYSTEMS WITH EMERGING TECHNOLOGIES: PREDICTIVE SIMULATION AND EMERGING TECHNOLOGIES

This critical infrastructure resilience research roadmap report focuses on better infrastructure modeling and simulation and leverage of future innovative technologies. It covers and extends the classical definition of a critical infrastructure system and emphasizes known interdependencies of such infrastructures, which leads to the question of better understanding interfaces and interdependencies. Simulation resources are proposed to be much extended using advanced computing approaches as well as more flexible and scalable modeling approaches that will also cover uncertainty modeling. Modeling and simulation are expected to be conducted in the future on various levels, with an ever-increasing fraction of empirical data-driven approaches. The gathering of data is assumed to be more and more automated using avionics approaches and specific as well as open-source data. Up to large-scale megacity real-time monitoring and simulation are expected to be realized including with respective decision and planning support actions. Standardization activities are recommended to be also supported by academia to ensure the consistency and take-up of already existing worldwide approaches. An open question was identified as to how to better frame critical infrastructure simulation approaches and existing solutions for installation within overall risk control and resilience assessment and improvement processes sufficiently up-taking results of research projects.

1.3.1 BACKGROUND AND INTRODUCTION

At the 2nd International Resilience Workshop (IRW) 2018 at Nanjing and Shanghai [1] working group 2 was originally assigned to address the following topic (see also the introductory overview to the workshop reports for detailed background and motivation) [2]: How to improve critical infrastructure systems and the emerging technologies.

This topic was detailed as "How to improve our capability of modeling and forecasting (including uncertainty quantifications of) such hazards and events with a better understanding of infrastructure, combined with advances in modeling and smart technologies that promise opportunities for groundbreaking discoveries to improve resilience. For instance, how to transform infrastructure, from physical structures to responsive systems, and to assure that increasingly interconnected CIS meet demands and withstand environmental hazards."

Hence, this working group addresses the question of how to improve the capability to model and forecast the effects of natural and extreme events on critical infrastructure systems, from immediate responses up to long-term socio-technical system behavior, including uncertainty quantifications. The approaches proposed will enable and leverage a better understanding of infrastructure, combined with advances in modeling and smart technologies that promise opportunities for groundbreaking discoveries to improve resilience.

Generic vistas include the definition of frameworks that comprise "loading" modeling, and socio-technical system modeling through all possible resilience response cycle phases and respective logic and time-ordered events. Such modeling and simulation frameworks need to take account of multiple possible system trajectories, including, e.g., recursions and self-reinforcing paths (snowball effects, cascading effects).

One important challenge includes understanding the interfaces, interdependencies and levels of independency between model components and how they are driven by natural, physical-engineering, technical, cyber, human, and social factors.

Within a socio-technical critical infrastructure system, modeling advanced approaches needs to cope with the challenge of multiple possible system trajectories of socio-technical infrastructure systems during and post disruptive events. The more the number and "length," i.e., time duration, number of logic steps, layers of assessment, of considered system trajectories are increasing, the more interesting resilience aspects are covered, including adapting, learning, and thus improving systems.

This challenge of being able to handle millions of possible representative events also needs to be addressed when assessing statistic, systemic, and deep uncertainties of risk control and resilience enhancement modeling, e.g., by using Monte Carlo approaches. It seems to be an open question whether predictive engineering modeling will ever master such ambitions, and, if at least in parts not, due to state and

Introduction

transition explosion, which high abstraction levels of modeling approaches could be sufficient for application purposes, e.g., abstract models like generic Markov models or probabilistic network models.

The current increment of extreme events and disasters all over the world due to climate change but also to increasing complexity and interdependency of modern communities (e.g., at urban level) highlights the fact that a policy for growth that will safeguard our medium and long-term prosperity must emphasize innovation far more than has previously been the case. This innovation also has to improve existing systems to keep them able to face new risks and multiple-hazards, but also lead to fundamentally new solutions and breakthroughs.

While the road to the enhancement of short-term, e.g., real-time, modeling, and prediction capabilities of critical infrastructure systems and subsystems in the case of disruption events through smart technologies such as interconnected sensors, mobile devices, scenario surveillance data, etc., seems rather straightforward, in particular as a continuous calibration and improvement of models already employed, one main challenge is to understand how to model and predict the more medium- and long-term effects of self-learning or even artificial intelligence (AI) systems on overall risk control and resilience of critical infrastructure systems. A further challenge is how to improve smart technologies such that they actually support the handling of major undesired events, as opposed to technology that focuses on optimizing systems close to standard operation.

The working group decided to address these challenges based on a flexible definition of critical infrastructure and infrastructure sector, starting with the identification of several main rather abstract objectives (Section 1.3.3). This is followed by the related main challenging topics (Section 1.3.4). Each topic is covered within subsequent sections, each covering major research gaps, hints of frameworks to address the challenges, and finally concepts, methods, and technologies to be further developed in the future.

In this way, Section 1.3.5 addresses interdependency modeling. Section 1.3.6 addresses the interlinked steps of sensing, modeling, and evaluation of results. Section 1.3.7 deals with future health monitoring and early warning systems. Section 1.3.8 discusses resilience-based improvements for critical infrastructural systems already in service. Section 1.3.9 presents the resilience-based design topic for infrastructure and systems, community resilience, and demonstration flagship projects (living labs). Section 1.3.10 is devoted to emerging technologies toward the innovation of resilient infrastructure and systems with reference to design and planning. The last sections provide a tentative roadmap (Section 1.3.11) and a summary in Section 1.3.12.

1.3.2 CRITICAL INFRASTRUCTURE DEFINITIONS AND MAIN OBJECTIVES LIST

"Resilience is the ability of a system to withstand a major shock within acceptable degradation that is recoverable in reasonable time, cost, and risk." This is the definition adopted by group 2. It applies well when critical infrastructures are considered. Earthquakes, tsunami, floods, explosions, impacts, hurricanes, and their combinations have been recognized as major disasters that can affect critical infrastructures.

Accordingly, with slight extensions of the definition of the USA PATRIOT Act [3] and focusing on the classes of critical civil infrastructures, they can be summarized in the following five main assemblies.

1.3.2.1 TI – Transportation Infrastructure
- Trucks, highways, and bridges
- Trains and rail tracks
- Airplanes and airports
- Ships and ports

1.3.2.2 EI – Energy Infrastructure
- Pipelines (including for green gas) and refineries
- Electrical grids, towers, and power stations
- Large-scale renewable energy generation and supply systems, e.g., offshore parks
- Nuclear reactors
- Dams

1.3.2.3 WW – Waste Water System
- Water pipes, tanks, and reservoirs
- Sewage conduits and refineries

1.3.2.4 ES – Emergency Services
- Hospitals
- Fire stations
- Police stations

1.3.2.5 IT – Information Technologies
- Sensors, connectors, and other data acquisition devices (DAQs)
- Interpretation of signals
- Databases and cloud computing
- Security and safety of data
- Artificial intelligence
- Machine and deep learning

1.3.3 BR – BUILDING STRUCTURES AND RESIDENCES RESISTING EXTREME LOADS

- Tall buildings
- Residences
- Heritage structures
- Industrial structures

Based on this bottom-up constructive definition of resilience and the main infrastructure elements that need to be considered, the following questions can serve as guidance for further identifying research needs regarding infrastructure systems

Introduction

modeling and prediction and dedicated related technologies, see also the question catalogs at the end of [2]:

- How to improve the identification of critical infrastructure, resulting in better system models?
- Which interdependencies of infrastructure elements should be considered?
- Which known and thinkable disruption types should be covered in the future?
- Which modeling and prediction quantities are of interest, which key performance indicators of infrastructure and service functions?
- What are the potential promising modeling and prediction frameworks, including for self-learning systems?
- What are the sufficient system models, e.g., covering all main system functions?
- What are future modeling and prediction ambitions, e.g., with respect to time, resolution, probability?
- How to relate modeling options to the aims of modeling and prediction?
- How to cover the coupling and interdependence of infrastructure domain models, of model hierarchies?
- How to model real humans and organizations from a "technical" perspective?
- What are future model-based simulative risk control and resilience optimization options taking into account expected methodological and computational advances?
- How to access and handle input data and information for ever-increasing models taking appropriate account of ownerships and interests (see also working group 3 report [4])?
- How will assessment capabilities affect insurance and investment models?
- How to take advantage of overall framing models (see also working group 1 report [5]) to better address real operator, citizen, and stakeholder needs by resilience research?

1.3.4 Key Challenges

The subsequent Sections 1.3.5 to 1.3.10 respectively cover one main future resilience research topic believed to be of dominating impact on successful research activities:

(1) Better understanding of inter and intra dependencies of critical infrastructure systems at various scales
(2) Advanced modeling and simulation, including human behavior
(3) Better simulation approaches to key challenges like progressive collapse for large sets of buildings of the real built environment
(4) Improvements for buildings and infrastructures in operation
(5) Efficient health monitoring and early warning with new technologies
(6) Improvements of existing infrastructures with new technologies
(7) Data-driven approaches
(8) Improvement of sensing, data acquisition, and management

(9) Leverage of machine learning and artificial intelligence technologies
(10) Evidence and science-based advancement of building codes and community resilience management

In each case, selected major research gaps, frameworks, and research contexts to address the challenges and new concepts, methods, and technologies to be further developed in the future are given.

1.3.5 Better Understanding Intra and Interdependencies of Critical Infrastructure

Understanding the impact of disasters on the civil infrastructure network allows the guiding of strategic pre-disaster hazard mitigation and post-disaster recovery planning of a community. However, civil infrastructures depend on each other to exchange products, information, or services. When disasters happen, these dependencies would aggravate the initial damage and lead to cascading failures. Thus, understanding the dependencies among infrastructure facilities is also essential in modeling the damage and recovery of a community under disruptive events.

Interdependencies can be described through a matrix approach that may also be useful for numerical implementation with the use of logic functions (e.g., in large-scale simulation models and within simple input-output models). Table 1.1 identifies the interdependencies between the six critical civil infrastructures of Section 1.3.3 that should be much further resolved in simulation approaches. For instance, WW does not influence TI ("No") but ES influences TI ("Yes").

When the model of a critical infrastructure is developed, it must comprise the systems itself but also the interaction with other infrastructural systems. This last requirement is probably the main issue to be solved. Indeed infrastructures are becoming more and more interoperable and interdependent within complex urban environments. Connected with this issue are the input requirements of modeling and simulation approaches that are affected by several parameters for describing the hazard itself and such as uncertainties, risks, actions, and their spatial distribution that may be difficult to define. This is more evident when stochastic inputs, e.g., earthquake or wind,

TABLE 1.1
Interdependency-Based Modeling for Critical Infrastructural Systems

Infrastructure	TI	EI	WW	ES	IT	BR
TI	Yes	No	No	No	Maybe	Yes
EI	Yes	Yes	No	Maybe	Maybe	Yes
WW	No	Yes	Yes	Yes	Maybe	Yes
ES	Yes	No	No	Yes	Yes	Maybe
IT	Maybe	Maybe	Maybe	Yes	Yes	Yes
BR	Maybe	No	No	Maybe	Maybe	Yes

Introduction 25

and multiple-hazards events, are considered. However, deterministic loadings are also characterized by several parameters and (systematic) uncertainties.

1.3.6 DATA-INFORMED AND DRIVEN INFRASTRUCTURE MODELING AND SIMULATION

Models may comprise numerical (e.g., FE – finite element or Applied Element – AE models) but also analytical ones. Besides, physical and field models can also be developed. Connected with this is the model validation that is usually a complex process to develop and conduct. This is more evident when community and multi-layer hybrid systems are considered. Indeed, it is usually difficult to have available validation data for complex systems, such as real data, from virtual cities and urban environments, or emergency evacuation due to extreme events, as in cases of explosions and seismic events. However, when input and output data are available, models can also be extrapolated, e.g., using machine learning or bio-inspired algorithms.

Connected with the emergency evacuation issues, e.g., by employing agent-based models, is human behavior (HB) modeling. Certainly, the role of emotions and altruism, for example, can be crucial during evacuation due to external shocks and may drive the decision making procedures.

An issue related to both the human behavior and modeling of critical infrastructures is the structural collapse. Indeed, it may affect the neighboring structures or interconnected critical infrastructures. Furthermore, local collapses or partial failures may affect the entire system collapse (progressive collapse) or its functions. It needs to be better covered since its correct resolution is critical for any relevant modeling approach, in particular, due to its binary effects on rather large scales.

Figure 1.1 orders some elements regarding sensing, simulation, and infrastructure improvements discussed above and in the following text sections.

Lessons from the past and historical disasters of critical infrastructure systems may allow for the establishment of post-disaster performance standards and objectives, e.g., for housing and reconstruction. Furthermore, it may also allow us to learn how to respond to the financial impact of losing a major portion of a customer base, to plan the economic recovery, and to evaluate possible insurance perspectives on disaster management.

Besides, the opportunity of developing large-scale numerical simulations and real-time hybrid simulations (sensing and numerical) is also a critical interest. Indeed, they allow for us to assess and measure how existing communities can respond to disasters and to plan possible countermeasures to improve the community response. Focusing on real-time hybrid simulations may play a special role during the evolution of the disaster effects to predict short-term dangerous situations and adopt optimal choices from the decision makers.

1.3.7 FUTURE HEALTH MONITORING AND EARLY WARNING

Connected to critical infrastructures and resilience is the development of sensing technologies that may support the decision makers and disaster management. Satellite technologies are gaining more and more interest in their ability to monitor

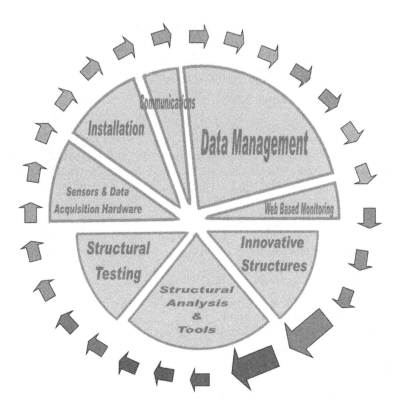

FIGURE 1.1 Sensing, simulation, and evaluation of resilience of critical infrastructures.

large-scale conditions and for the possibility to predict extreme weather events. This is also driven by free access to high-quality data.

Early warning systems (EWSs) against natural hazards have focused the attention of scientists and designers in the last decades on creating an effective tool for improving the resilience of communities and systems. EWSs can provide a few to a few tens of seconds warning prior to damaging ground shaking and are currently operational in Mexico, Taiwan, and Japan. Their use is connected to the recent progress in sensing systems technology, remote sensing, and wireless systems.

The adoption of smart devices in emergency environments is becoming more and more important. Several scenarios, such as post-earthquake and fire emergency activities, are very attractive for possible applications, even if they present several scientific challenges that must be addressed in order to satisfy rescuers' requirements. The network of modular programmable sensors (anchors) equipped with specific sensing units (e.g., acceleration, humidity, temperature) can be used to track the agent (e.g., the fireman) in the dangerous scenario and for monitoring its current state safely (e.g., heartbeat, breathing). Therefore, the system may monitor both the agent and the environment conditions in real-time.

Critical infrastructures and urban structures need constant maintenance and inspection of the structural health conditions and safety of the users, however, to access the structure is getting harder and harder due to their enormous height and size. In order to deal with this problem, many researchers have developed robots for system health monitoring (SHM) inspections. Drones include, e.g., climbers, flying objects, and much improved remote mobile cameras. They may play a critical role for high-level inspection and, thus, for maintenance.

1.3.8 RESILIENCE-BASED IMPROVEMENTS FOR CRITICAL INFRASTRUCTURAL SYSTEMS ALREADY IN SERVICE

For reasonably expected and largely expected events to occur during the service life of a structural or infrastructural system may be useful to plan retrofitting, renovating, and repairing actions. To this aim and to withstand the external hazard while protecting both new and existing structures and infrastructures, passive control (e.g., base isolation), active, semi-active, and adaptive control systems can be useful. They can also be designed for special structures in infrastructural systems (e.g., long-span bridges) as a protection to multi-hazard conditions (e.g., wind and earthquake). Furthermore, control systems can be used to improve system resilience through the automatic compensation of possible out-of-service structural components (immediate resilience). Research challenges include how to take advantage of existing and emerging solutions for much larger fields of application, also by reducing the investment costs.

Focusing on structural and infrastructural renovation, duplication of critical components, or functions of a system with the intention of increasing reliability of the system, usually in the form of a back-up or fail-safe, or to improve actual system performance, can be essential to provide additional reserve (redundancy). It means that the failure of a component does not result in the collapse of the entire system, and alternative loading paths can be provided. This concept may play a significant role for existing structures but also for the design of new ones, and could be included in new standards and guidelines. It remains challenging to field the most efficient solutions while taking account of real-world constraints in a more systematic way.

1.3.9 RESILIENCE-BASED DESIGN FOR INFRASTRUCTURE AND SYSTEMS, COMMUNITY RESILIENCE, AND DEMONSTRATION FLAGSHIP PROJECTS (LIVING LABS)

A need for resilience-based design, planning, and optimization is arising with reference to infrastructure systems. Indeed, innovative approaches to decision making methods for the design of new infrastructures in times of climate change, multi-hazard conditions, and increasing interdependencies are expected. The result directly downstream of this innovation process is the creation of new guidelines that are directly available and can lead to a general development in the direction of the creation of resilient communities. These are new standards that need to include new design clauses, technologies, and specifications for new and existing structures.

Besides, community preparedness is also a crucial aspect toward resilient communities and can be deployed on many levels, from the higher-education level (e.g., engineers) to the training of technicians and citizens. If the first one is mostly developed at the academic level, citizens may be prepared to withstand and cope with disasters through civil protection and emergency agencies.

Demonstration projects can also be considered at this stage, e.g., Following the devastating 2004 tsunami, the development of the Indian Ocean Tsunami Warning and Mitigation System [6] was initiated at the World Conference for Disaster Reduction in 2005 under the lead of the United Nations Education Scientific and Cultural Organization's Intergovernmental Oceanographic Commission. From that point, many organizations have been engaged in the task of developing tsunami early warning systems and community-based disaster risk management in coastal regions.

1.3.10 Emerging Technologies for Innovation of Resilient Infrastructure and Systems in Design and Planning

Rapidly increasing urbanization is associated with many challenges that need to be addressed. Emerging technologies are expected to help with improving community resilience and infrastructures supporting urban disaster risk management. Climate change and urbanization are increasing the risk and impact of disasters and rapid urban development has been driving up urban risk. Mega-disasters are happening more frequently, and so-called everyday crises and other stresses are heightening vulnerability and undermining coping capacities. This, coupled with the growing urban populations, makes it critical for organizations to better support community resilience so that people living in urban areas can help themselves, as frequent shocks and stresses become a common part of everyday life.

This leads to the concept of developing smart cities with strengthened infrastructures and improved quality of life.

The use of smart materials, nanoscience, and nanotechnology obviously arises as the natural choice for attaining such objectives. On this basis, emphasis must be made on the requirement of active engagement of scientific research and engineering applications in the area of smart materials and nanotechnologies for future cities; e.g., the addition of nanocomposite materials into cement has shown notable potential in improving its performance and compressive strength.

In this light, traditional disciplines, such as SHM and control, are required to make a new effort to consider and understand the new and evolving conditions. In addition, it is also necessary to push in the direction of new, multidisciplinary solutions, such as civionics, which, like avionics and mechatronics, still need to be fully understood and developed, see Figure 1.2. The obvious intrinsic difficulties for the development of civionics are to take advantage of scale factors and forces as are

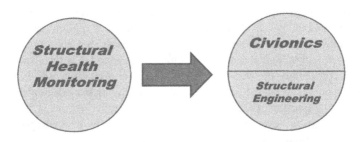

FIGURE 1.2 Expected development from classical system health monitoring to advanced approaches.

typical in civil engineering, e.g., to provide infrastructure operators with cheap and safe monitoring devices.

Implementation of existing sensors and the development of new sensors for civil engineering needs (e.g., fatigue sensor, fire sensor) toward real-time monitoring of civil structures and infrastructures is one of the primary actions that is expected. The challenge is to launch existing technologies at the research level toward applications in the real-world.

Data acquisition systems (DAQ or DAS – sensors, to convert physical parameters to electrical signals; signal conditioning circuitry, to convert sensor signals into a form that can be converted into digital values; analog-to-digital converters, to convert conditioned sensor signals into digital values) play a major role in SHM, and they may also be critical components in developing resilient communities and infrastructures.

Developing equipment that is waterproof, robust, and reliable, able to survive in harsh construction processes, is also a foremost requirement toward new and effective solutions. Efforts toward simplification and reliable implementation include the development of repeatable (regarding ad hoc built-up) wireless systems.

Also, the provision of cheap localization of non-rescue persons and objects indoors remains challenging. Such a capability would open up many possibilities, e.g., much better priority-driven rescue coordination.

As a result of an increasingly monitored world, an exponential increase in the data collected should be expected. Therefore, data mining, management, processing, and interpretation will require major research and development efforts at the university and industrial level. Connected with this is the development of web platform bases and cloud monitoring solutions, as part of an emerging process to evaluate, monitor, and manage cloud-based services, applications, and infrastructures.

Technologies like augmented reality in construction are emerging to digitalize the construction industry, making it significantly more effective. Furthermore, the digital twin – a concept of having a real-time digital representation of a physical object – is also an emerging technology toward resilient communities and infrastructures. Digital data is formed from sensors that continuously monitor changes in the environment and report back the updated state in the form of measurements and pictures.

Big data and data mining play an important role in an increasingly complex world, as well as machine learning, deep learning, and bio-inspired algorithms. Indeed, these last-mentioned technologies have changed the old paradigm "input-algorithm-output" toward a new scientific creativity in a wide range of fields and applications.

1.3.11 ROADMAPS AND STRATEGIES PROPOSED FOR FUTURE IMPLEMENTATION

The following goals can be aligned into an ambitious timeline, with appropriate measures along all phases (short term, medium term, long term):

- Refinement of infrastructure taxonomies, ontologies, and definitions along with key functionalities and interdependencies.
- Provision of fast reference models for individual infrastructure systems, including interfacing models.

- Provision of fast algorithms over various levels of rigor and modeling refinement for such reference models, e.g., abstract, engineering-parameterizing, system-analytic to engineering-physical-simulative.
- Development of interaction and uncertainty assessment models and simulations and well-founded fundamental theories.
- Including uncertainty modeling at all levels.
- Advancement of single key approaches and technologies, including:
  - Better understanding of tipping-point damage modeling, e.g., collapse, cascade initiation, emergent unintended behavior
  - Technologies for the safety of individuals, e.g., localization, orientation, personal protection
  - Large-scale natural hazard countering, e.g., urban fires, forest fires, countering draft events with natural approaches supported by technology
  - Advanced material with resilience built-in and green properties
  - Overall green life-cycle resilience
  - Avionics for advanced in-situ and efficient and safe air-based monitoring
  - Leveraging the "sensor dust" idea into reality
  - Taking advantage of the ever better sensors in smartphones, smartwatches, and further mobile devices
- AI and machine-learning-driven active protection systems with no material or physical redundancy, i.e., with nothing but intelligent system immediate response resilience back-up.
- Massive actor systems that can project or allocate almost any resources in a short time are needed to protect, respond, rebuild, or adopt successfully, rather than generating massive local redundancies.

It is expected that future implementations of technological and scientific advances toward resilience improvements will be made through research results and also by the development of new guidelines and standards. Indeed, although it is vital to continue scientific research and develop new technologies, and therefore advance the various fields that can support new infrastructures and resilient communities, it is essential to create a strong link between research and scientific application. This link, obviously adapted to real-life conditions and local social, economic, and political issues, is represented by guidelines and standards. In other words, the key rules are established by governments themselves in order to unify and improve the conditions of their countries. Here, supporting scientists and country authorities should take up existing approaches and collaborate to further pool their expertise.

1.3.12 Summary

Based on a detailed definition of the resilience of critical infrastructure, infrastructures at risk, and the assessment of expected interdependencies, the present report delivers key working areas (see Sections 1.3.5 to 10 for future resilience as well as (additional) appropriate approaches (see Section 1.3.11)). Many of the proposed future innovations are envisioned to be ready for operational use on mid-term time scales or even shorter, especially in the area of automated air-born inspection and rescue forces support. Further envisioned research activities are recommended to cover such areas as leveraging computational parallel graphical processor approaches for large-scale infrastructure models and simulation and taking advantage of dedicated sensor and open-source data. Real-time capabilities need to be developed to better support operators, responders, and infrastructure users.

Thus working group 2 analyzed how to improve critical infrastructure systems by using the emerging technologies. The main critical infrastructures and their interdependencies were identified. Modeling and simulation issues were deepened with the emerging technologies related to monitoring and early warning systems. These two are the two sides of the same coin, the virtual simulation environment and the systems that may collect the data in real-time from the real-world. Both can be implemented separately or integrated into a hybrid framework. The current existing application and the emerging technologies are also discussed as lessons learned and future visions for resilience infrastructures. Finally, the need for standard and guideline developments for a comprehensive implementation of resilience was also identified as a strategic roadmap for the future.

LITERATURE

1. *2nd International Resilience Workshop (IRW) 2018*, Nanjing and Shanghai, China, October 31–November 2, 2018, http://www.workshop-china2018.resiltronics.org/, last access on November 23, 2018.
2. Lili Xie, Gian-Paolo Cimellaro, Michel Bruneau, Zhishen Wu, Max Didier, Mohammad Noori, Aftab Mufti, Xilin Lu, Jinpin Ou, Shamim Sheikh, Ying Zhou, Marco Domaneschi, Teruhiko Yoda, Ertugrul Taciroglu, Ivo Häring, Anastasios Sextos. (2018). Challenges and generic research questions for future research on resilience of critical infrastructure: introduction to workshop reports of the 2nd International Workshop on Resilience (IRW) 2018. *Proceedings of the 2nd International Resilience Workshop (IRW) 2018*, Nanjing and Shanghai, China.
3. US Public Law Pub.L. 107–156, https://www.gpo.gov/fdsys/pkg/PLAW-107publ56/html/PLAW-107publ56.htm, last access on November 18, 2018.
4. Lili Xie, Gian-Paolo Cimellaro, Michel Bruneau, Zhishen Wu, Max Didier, Mohammad Noori, Ivo Häring. (2018). Frameworks, fundamentals and education for future infrastructure risk control and resilience: workshop 1 report. *Proceedings of 2nd International Workshop on Resilience (IRW) 2018*, Nanjing and Shanghai, China.
5. Teruhiko Yoda, Ertugrul Taciroglu, Ivo Häring, Anastasios Sextos, Mohammad Noori. (2018). Big data analytics, machine learning and AI for future human support against disruption events and for critical infrastructure system resilience: report of working group 3. *Proceedings of 2nd International Workshop on Resilience (IRW) 2018*, Nanjing and Shanghai, China.
6. Indian Ocean Tsunami Warning and Mitigation System, http://www.ioc-tsunami.org/index.php?option=com_content&view=article&id=8&Itemid=58&lang=en, last access on November 27, 2018.

1.4 BIG DATA, MACHINE LEARNING AND AI FOR FUTURE HUMAN SUPPORT IN DISRUPTION EVENTS AND CRITICAL INFRASTRUCTURE SYSTEM RESILIENCE

Recent research results as well as lessons learned after man-made and natural disasters reveal that critical infrastructure relies on interactions between humans and technical solutions that are implemented during a crisis. Such experiences and the continuously increasing amount of available data, analytics, and intelligence stimulates the question of how future resilience research can exploit technological advancements to improve the capacity of both humans and critical infrastructures during disruption events. This section which outlines a summary of a workshop report summarizes the options available to (a) involve users, operators, and decision makers in joint research, (b) take advantage of digital semantic urban and rural data, to use machine learning to determine input parameters for the modeling and simulation of infrastructures, (c) design a modular hub for storage of information and risk and resilience assessment with respect to a broad set of potential threats, as well as (d) use of similar modules within systematic procedural approaches, such as, spatial scenario definitions, person exposure, abstract threat visualization, hazard, damage and risk computation modules, visualization of damage, frequency, risks, and resilience quantities along with their evaluation involving all stakeholders. It is envisioned that the scientific community can build around a hub platform with the aim to enhance the resilience of critical infrastructure networks in terms of downtime and disruption costs, including large-scale, "what-if," vulnerability scenarios. This section also provides details regarding motivation, challenges, and key research approaches, toward the "all-resilient" systems of the future.

1.4.1 MOTIVATION, BACKGROUND AND INTRODUCTION

When inspecting the current critical infrastructure resilience research landscape from a technical science outlook, consensus of the overall ambition has been reached in terms of an advanced formulation of overall risk management: the individual, socio-economic, and environmental losses before, during, and after disruptions of critical infrastructure due to natural and man-made disruption events need to be rigorously assessed and reduced to an acceptable level.

This covers losses in single resilience cycle management phases as well as overall losses. Disruption events can be massive and sudden or creeping and gradual. Losses can be expressed in terms of expected direct and indirect consequences in the form of damage or loss quantities, commonly expressed in a probabilistic manner.

Even though the above abstract notion of the resilience objectives is generally agreed, the challenge to reach such goals is still debatable among different schools of thought originating from different disciplines such as engineering, economics, social sciences, and humanities, the latter often being rather skeptical of "technical solutions." As a result, the resilience of critical infrastructures remains a versatile task, in particular when approached with mainly engineering, technical, and natural science research instruments while taking into account the social, societal, individual, and psychological perspectives.

Benefiting from the engagement of the participants during the 2nd International Resilience Workshop (IRW) 2018 that took place in Nanjing and Shanghai, China, between October 31–November 2, 2018, working group 3 focused on the improvement of the resilience of critical infrastructures by exploiting recent technological advancements, particularly in terms of big data acquisition management and processing, modeling, AI, and other novel approaches (see also the introduction to the workshop reports [1] for reference to working group 1 that covers conceptual aspects [2] and working group 2 that discusses research needs to advance modeling and simulation of critical infrastructure along with the related technologies needed for data, processing, and the implementation of approaches [3]).

Taking advantage of the existing and future options offered by big data analytics, machine learning, and artificial intelligence for human support in cases of disruption events within critical infrastructures, the main technological and engineering challenges are critically discussed in light of addressing the following questions on how to:

- Enhance societal preparedness and societal resilience against the impacts of natural and man-made hazards.
- Make sound research investments to better develop technology that supports critical infrastructure and human–technology interactions.
- Promote a multidisciplinary collaboration between the engineering, ICT, social sciences, economics, and humanities disciplines.

From a systemic perspective, the resilience of individual assets to single threats is rather well understood. This holds, in particular, when noting that most infrastructure subsystems when sufficiently isolated can be well assessed almost to any degree of resolution and with respect to all aspects when resorting to traditional disciplinary sciences, e.g., classical earthquake engineering, by doing more and better "of the same or similar" approaches. However, there is still a lack of a holistic framework to assess resilience at an infrastructure system level considering all interactions among individual components, networks and inter-system levels.

At the same time, it was acknowledged that new scientific approaches on a subsystem and component level are critical for overall system resilience. However, it was argued that only in the context of the overall system assessment can it be decided whether such resilience improvements are of real benefit. This is an argument for avoiding allocating resources to non-relevant system capabilities and designs for modeling and simulation and even more in real-world rather than into real bottlenecks for better resilience.

Looking at the challenging nature of the interlinked systems exposed to an increasing (multi-event) threat landscape, there is a lack of transdisciplinary research and direct collaboration to quantify and improve resilience involving developers, operators, and users (DOU) including but not limited to users, citizens, engineers, risk analysts, social scientists, economists, stakeholders, and decision makers, e.g., in terms of co-creative activities and citizen-involving research.

As modern communities become more interdependent than ever, the potential impacts of hazardous events have a broader potential footprint on infrastructure systems. The frequency, modality, and level of extremity of known and new hazards (e.g., climate, cyber-related, "AI-related"), as well as our exposure to them, is expected and often (already) observed to be increasing massively. At the same time serviceability, functionality, safety, and security (technical) capabilities of infrastructure systems are increasing massively as well. In particular, technological capabilities and the ability to generate, harvest, predict, and process data and relevant system information are rapidly enhancing. Generally, the question arises of whether the potential weaknesses of modern infrastructure systems can be cured with technological means. This will be one of the guiding questions and will be shown to be answered rather optimistically in the present report.

Resilience objectives compete with economically driven factors that affect the desirable functionalities. Such competing demands and optimization goals are raised along with different paradigms, such as lean production, just in time, slim systems, avoidance of cold standby, one-stop-shop solutions, outsourcing of infrastructure of business, etc. Advanced risk control and resilience furthermore have to compete with more increasing sustainability aims, in particular, with the strong need to reduce the carbon footprint of critical infrastructures.

The report continues with a list of objectives for developing resilient systems with a focus on data and self-learning technologies, see Section 1.4.2. Section 1.4.3 identifies key challenges, which are believed to be feasible and ambitious research targets for future critical infrastructure research. Section 1.4.4 gives an overview of current research gaps as identified by working group 3. Section 5 presents framing assumptions, processes, and development lines on how to systematically address the identified research needs. Section 1.4.6 discusses promising approaches, methods, and technologies in more detail. Section 1.4.7 roadmaps the proposed approaches. The overall conclusions of the working group are delivered in Section 1.4.8.

1.4.2 Objectives

Taking into account the background given in the introduction and being aware of such drivers as man-made, anthropogenic, natech, accidental, sabotage, criminal, human-malicious terrorist events affecting critical infrastructure resilience and related research and implementation ambitions, the following generic objectives were identified, most of which build on each other:

1. Provide a methodology to develop valid and specific frameworks for resilience assessment and generation.
2. Define and quantify a hierarchy of resilience objectives from the asset level to systems, systems of systems, and community level, possibly even further to the nation, worldwide region, and worldwide level.
3. Build coupled socio-cyber-physical models, methods, data, and tools to assess and to optimize resilience.
4. Develop tools and interfaces in a modular and generic way that can be easily adjusted for new hazards.
5. Provide opportunities for collaboration between DOUs and promote adoption by end-users.
6. Better understand the loading and threat vector as well as the interdependency of different hazards and quantify the local and global footprint of each hazard.
7. Understand, model, and improve our models for the rapid recovery of components and systems.

Introduction

8. Demonstrate and validate the importance of system-level resilience assessment and investments for the improvement of socio-economic prosperity and well-being.
9. Better understand how to quantify resilience objectives; in particular, to be able to quantify them and discuss them with more established systemic goals such as efficiency, profitability, and ecologic sustainability on various scales.
10. Take advantage of advanced approaches such as (big) data and related technologies like machine learning, statistical inference, etc.
11. Efforts in these areas should always keep sight of the big picture.

1.4.3 Key Challenges

Key challenges identified that could be addressed with promising results in future resilience research include but are not limited to:

1. Lack of common language and approaches.
2. Lack of will to collaborate across the boundaries of different disciplines to overcome school of thought thinking and disciplinary silo approaches.
3. Lack of more systematic (versus just presenting what works) selection and leveraging of systemic approaches from complex system theory, cyber-physical systems modeling, graph models, Markov models, etc.
4. Fragmentation and/or inaccessibility of data with unknown levels of reliability (bad quality data).
5. Big data may not be so big, in particular, when assessing rare disruption events.
6. High computational cost and modeling challenges for interlinked, non-linear, time-dependent problems.
7. Epistemic (systematic) and aleatoric (statistical) uncertainties at various levels.
8. Liability concerns of stakeholders.
9. Exposure and responsibility with respect to threats is diffused.
10. Lack of senior leadership and barriers from within the (fragmented) communities.
11. Resilience research is hard to validate; there is a long (real-world) learning curve.
12. Strategy and response to natural and man-made hazards are cultural, financial, and experience-dependent, often resulting in ergodic and moving-target expectations and goals.
13. Hazard quantification is not globally (worldwide) uniform: different hazards are different in essence or are assessed differently. Even in cases where, for instance, quantitative performance-based results are comparable, with the help of unbiased scientific methods, using individual, (non) local, profile, group, or individual personal and objects risks (on resilience objectives), the risk and resilience evaluation (risk and resilience appetite) can be ultimately quite different.
14. Resilience is not yet generally believed and/or known to be a business case for success.

1.4.4 MAJOR GAPS IN THE STATE OF THE ART

In the years after 9/11, several threads of research can be identified regarding critical infrastructure contextualization, assessment, and improvement. Regarding the level of detail of investigations, at least in the European Union (EU) research calls opened by the European Commission (EC), a business-ready technology is sought with strong relevance to the civil security user community as well as for massive company involvement toward good economic prospects.

Regarding the infrastructure application domain, it can be observed that the initial enthusiasm for generic approaches with a high relevance for all critical infrastructure operators is somewhat decreasing. When referring again to EU calls, the expected use case scenarios to be addressed by research are developed for most calls in detail, consulting expertise gathered from a wide and by now rather established spectrum of stakeholders representing diverse interest groups. There is currently a consensus that critical infrastructure resilience should be driven mainly by domain-specific operators, rather than asking for generic solutions that are applicable to all types of infrastructures. For instance, this became apparent within a call that asked for the development of generic approaches to EU critical infrastructure resilience, which was answered very differently depending on the infrastructure use cases in a series of EU research projects, see, e.g., the projects RESILENS [4], RESOLUTE [5], and DARWIN [6]. Other calls ask from the very beginning for infrastructure-specific sectorial cyber-physical risk control (see e.g., the call on prevention, detection, response, and mitigation of combined physical and cyber threats to critical infrastructure in Europe) [7]. This can be motivated by the insight that resilience assessment and improvement is most efficiently conducted at the level of economic decision makers by addressing improvements possible without changing the legal and societal framework. The research gap is how to identify and deliver relevant evidence that requires users and decision makers to more adequately attend to future infrastructure risk control and resilience.

In this context, game-changing regulatory and legal modifications should be investigated to foster the decisions of operators, commercial and private critical infrastructure users, e.g., using serious gaming environments and simulations of CIS. However, research and frameworks that take into account systematically existing and potential boundary conditions and policies are missing. In particular, when asking for such frameworks that ask for the connection of policies to quantitative infrastructure models.

The last assessment can be understood as a specific gap that originates from the main research gap which is that there are not yet country-wide, multi-country, or even worldwide generally accepted approaches and even fewer standards to understand, model, and simulate interconnected critical infrastructure systems as well as single infrastructures. If domain-specific standardization approaches have been successful, they have been supported by research insights. However, most such standards are only available on the community level. There is a lack of further such efforts covering resilience from a more generic technical perspective and much more so for specific infrastructure domains.

Of particular interest are infrastructure static and dynamic data, as functional (non)performance (service) data that is necessary and sufficient for understanding,

Introduction

modeling, and simulating potential snowball effects, inter and intra cascades of potential events. Coupled infrastructure systems need to be understood well beyond a minimum description to determine such information and data bags.

Regarding research needs, methodological frameworks are missing that are capable of improving the efficiency of resilience approaches; the time scales of adaption to the true needs of society, economy, and the environment. For instance, it is challenging to identify motivating factors that lead to the increased engagement of actual decision makers, which are typically driven by economic revenue, branding options, or patented innovation.

Besides these more generic red threads and major gaps of resilience research, more specific research gaps can be identified that result in the research questions and key challenges identified in Sections 1.4.2 and 1.4.3. Gaps in current research include:

1. How to better cope with a lack of data, bad data, and big data.
2. How to address the lack of manpower that can be assigned to more frequent resilience issues with the help of more automated approaches such as those that are accessible with machine learning and AI approaches, in particular, for sensing, inspection, and pre-decision making.
3. How to design hybrid and partly autonomous systems for coping with rare events, which by definition generates little training data.
4. How to quantify resilience gain in terms of economic profit.

1.4.5 Framework to Address the Challenges

The frameworks of (semi)quantification, assessment, design, and/or development of resilient socio-technical systems include:

1. Risk management and analysis approaches that focus on "risks on expected resilience behavior" and respective resilience optimization and risk minimization problems (see, e.g., Section 3 of [8]). Of special interest are identified risks that need to be mitigated and for which methods without machine learning approaches do not (yet) sufficiently deliver.
2. Systematic system development processes such as the V-model with extensions or further developed spiral models. Of special interest are additional development steps (e.g., to determine resilience requirements) that require analysis that cannot be conducted with traditional methods.
3. Agile short-term and long-term processes driven by disruption events, successful mitigations in similar cases, and expectedly successful approaches, i.e., resilience improvement in a case by case approaches.
4. System analysis and simulation processes including, e.g., identification of resilience investigation targets, system understanding and modeling, system simulation or analysis, results discussion, assessment and evaluation, and executive recommendations.
5. Semi-quantitative expert and citizen opinion assessment and evaluation frameworks (see, e.g., Section 3 and 7 of [8]).

6. MCEER's standard and the PEOPLES framework, see respectively [9], as well a more recent framework for resilience quantification [10].
7. Performance function-based resilience assessment and improvement processes: Variations of system-function based risk quantification and improvement approaches that identify a system's main functions of interest and potential disruptions to determine critical combinations that require further investigations in terms of resilience quantification and assessment. Overall, resilience quantities are evaluated and in cases where they are not acceptable resilience improvement measures are selected. The objective is to obtain an overall sufficiently resolved and trusted system resilience quantification to show that all developed and implemented improvement measures lead to overall acceptable system resilience. (see e.g., [11]).
8. Another example is the structured approach of the "Foresight review of resilience engineering: Designing for the expected and unexpected" stipulated by Lloyd's register foundation, which provides background, definitions, and challenges while focusing on engineering solutions. It covers a rather wide range of fields [12].

1.4.6 Proposed Actions to Tackle the Challenges: Concepts, Methods, and Technologies for Future Resilience Research

The following actions are proposed to address the challenges, along with concepts, methods, and technologies believed to strongly support future resilience research:

1. Explore synergies between interdisciplinary groups and define common procedures and protocols for sharing data, methods, and tools.
2. Build more reliable data and metadata: ontologies, taxonomies, etc.
3. Take advantage of data analytics and machine learning techniques to overcome communication barriers and costs.
4. Extend risk management and analysis approaches to explicitly cover the objective of being resilient and for assessing "risks on resilience."
5. Apply consistent system functional approaches, system performance function quantification, including aggregated and overall resilience quantification.
6. Aim at "all-resilience" approaches.
7. Define data, modeling, and interfacing modeling and simulation hubs, extending and joining similar such environments, e.g.,
   (i) A Simcenter automation simulation platform [13]
   (ii) An Open Framework for Integrated Multi-platform Simulations for Structural Resilience [14]
8. Extend or combine existing simulation urban risk control and resilience frameworks as developed in the EU projects D-BOX, ENCOUNTER, VITRUV, and EDEN, see [15] and [16] for multiple potential terror events and other types of events as well as for single events, see, e.g., a simulation environment for terroristic explosive events [17].

Introduction

9. Develop frameworks for resilience-based design and assessment of structures *and* systems for life-cycle operation.
10. Develop data harvesting tools for building inventories (data pipelines).
11. Customize machine learning techniques for resilience applications.
12. Train public data sets for model calibration and verification.
13. Develop APIs and interface technologies between assets, sensors, geospatial data (GIS), smart systems representations (digital twins), digital building, infrastructure, and urban areas models, such as BIM, City-GML, Internet of Infrastructure (IoI), and early warning systems.
14. Enhance rapid real-time post-event capabilities.
15. Promote a seamless transition between system health monitoring and functionality assessment as well as for automated warning, disruption detection, monitoring, and recovery activities coordination (one-stop-shop to ensure regular use of platforms and tools).
16. Define core professional competencies for engineering practitioners and students in the domain of resilience concepts, design, development, audition, insurance, and research.

1.4.7 ROADMAPS AND STRATEGIES PROPOSED FOR FUTURE IMPLEMENTATION

Figure 1.1 presents a tentative roadmap for future resilience research focusing on the development of a worldwide hub for representation, modeling, and simulation of single critical infrastructure elements as well as networks and coupled networks. It is envisioned that such a development is conducted in several steps. It is believed that the scientific exchange of a potential user community of such a critical infrastructure modeling and simulation hub is crucial for success; one such opportunity could be a third international workshop on critical infrastructure resilience (see Figure 1.3).

1.4.8 SUMMARY AND CONCLUSIONS

It is expected that the above-described processes, methods, and technological advances will lead to scaling effects for resilience research capabilities, i.e., non-linear improvements. Especially when putting together semi-automated approaches on a science platform hub. Different participants could provide different parts of modeling and assessment capabilities for assessment and optimal improvement identification of critical infrastructure assets and systems.

This also positively answers the question of whether it is expected that scientific methods can be used to overcome risks as generated by ever more complex, interlinked, and smart infrastructure systems.

Such an envisioned platform would:

- Leverage the potential of using advanced digital data-driven and self-learning methods (AI approaches) for improving resilience, in particular, for data acquisition of diverse sources
- Facilitate a better assessment, quantification of risk, and resilience for large-area systems

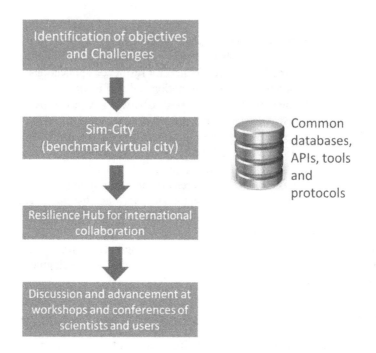

FIGURE 1.3 Tentative roadmap to an international exchange platform for large-scale area and critical infrastructure simulation and assessment, also showing how it can be iteratively improved.

- Take into account various scales, complexities, and interdisciplinarities
- Make standardized computations accessible for assessing fatalities as well as socio-economic and environmental losses
- Provide international assessment criteria for natural and man-made events
- Enhance societal preparedness
- Enhance societal resilience
- Explore counter natural and man-made disasters
- Provide technology that supports critical infrastructure operators
- Allow to better understand large-scale (on citizen crowd level) human–technology interaction
- Promote transdisciplinary collaboration
- Set the platform for a broad, multidisciplinary discussion on socio-political and policy issues that relate to the resilience of critical infrastructures to major disruptive events

REFERENCES

1. Lili Xie, Gian-Paolo Cimellaro, Michel Bruneau, Zhishen Wu, Max Didier, Mohammad Noori, Aftab Mufti, Xilin Lu, Jinpin Ou, Shamim Sheikh, Ying Zhou, Marco Domaneschi, Teruhiko Yoda, Ertugrul Taciroglu, Ivo Häring, Anastasios Sextos. (2018). Challenges and generic research questions for future research on resilience

of critical infrastructure: introduction to workshop reports of the 2nd International Workshop on Resilience (IRW) 2018. *Proceedings of the 2-nd International Resilience Workshop (IRW) 2018*, Nanjing and Shanghai, China.
2. Lili Xie, Gian-Paolo Cimellaro, Michel Bruneau, Zhishen Wu, Max Didier, Mohammad Noori, Ivo Häring. (2018). Frameworks, fundamentals and education for future infrastructure risk control and resilience: workshop 1 report. *Proceedings of 2nd International Workshop on Resilience (IRW) 2018*, Nanjing and Shanghai, China.
3. Aftab Mufti, Xilin Lu, Jinpin Ou, Shamim Sheikh, Ying Zhou, Marco Domaneschi, Mohammad Noori, Ivo Häring. (2018). How to improve critical infrastructure systems with emerging technologies: future critical infrastructure systems predictive simulation and emerging technologies – Report of working group 2. *Proceedings of 2nd International Workshop on Resilience (IRW) 2018*, Nanjing and Shanghai, China.
4. RESILENS, https://cordis.europa.eu/project/rcn/194842_de.html, last access on November 29, 2018.
5. RESOLUTE, https://cordis.europa.eu/project/rcn/194870_en.html, last access on November 29, 2018.
6. DARWIN, https://cordis.europa.eu/project/rcn/194846_en.html, last access on November 29, 2018.
7. EU secure union call on cyber-physical security of critical infrastructure, SU-INFRA01-2018-2019-2020, http://ec.europa.eu/research/participants/portal/desktop/en/opportunities/h2020/topics/su-infra01-2018-2019-2020.html, last access on November 30, 2018.
8. Ivo Häring, Stefan Ebenhöch, Alexander Stolz. (2016). Quantifying resilience for resilience engineering of socio technical systems. *Eur J Secur Res*, Volume 1, Issue 1, pp. 21–58. doi:10.1007/s41125-015-0001-x.
9. Chris S. Renscher, Amy E. Frazier, Lucy A. Arendt, Gian-Paolo Cimellaro, Andrei M. Reinhorn, Michel Bruneau. (October 8, 2010). A framework for defining and measuring resilience at the community scale: the PEOPLES resilience framework, Technical Report MCEER-10-0006. http://mceer.buffalo.edu/publications/catalog/reports/A-Framework-for-Defining-and-Measuring-Resilience-at-the-Community-Scale-The-PEOPLES-Resilience-Framework-MCEER-10-0006.html, last access on November 25, 2018.
10. I. Koilanitis, A. G. Sextos. (2018). Integrated seismic risk and resilience assessment of roadway networks in earthquake prone areas. *Bull. Earthqu. Eng.* (publication online). doi:10.1007/s10518-018-0457-y.
11. I. Häring, G. Sansavini, E. Bellini, N. Martyn, T. Kovalenko, M. Kitsak, G. Vogelbacher, K. Ross, U. Bergerhausen, K. Barker, I. Linkov. (2017). Towards a generic resilience management, quantification and development process: general definitions, requirements, methods, techniques and measures, and case studies. In: *Resilience and Risk: Methods and Application in Environment, Cyber and Social Domains*, Editors: I. Linkov, J. M. Palma-Oliveira, pp. 21–80, Springer, ISBN 9789402411225, http://www.springer.com/de/book/9789402411225.
12. Lloyd's Register Foundation Report Series: No. 2015.2, Foresight review of resilience engineering: designing for the expected and unexpected, October 2015, http://www.lrfoundation.org.uk/publications/resilience-engineering.aspx, last access on November 25, 2018.
13. Simcenter, Siemens, https://www.plm.automation.siemens.com/global/de/products/simcenter; last access on November 29, 2018.
14. UT SIM Hub, https://www.ut-sim.ca, last access on November 29, 2018.
15. G. Vogelbacher, I. Häring, K. Fischer, W. Riedel. Empirical susceptibility, vulnerability and risk analysis for resilience enhancement of urban areas to terrorist events. doi:10.1007/s41125-016-0009-x.

16. K. Fischer, S. Hiermaier, W. Riedel, I. Häring. (2018). Morphology dependent assessment of resilience for urban areas. *Sustainability*, Volume 10, Issue 6, p. 1800. doi:10.3390/su10061800.
17. I. Häring, M. von Ramin, A. Stottmeister, J. Schäfer, G. Vogelbacher, B. Brombacher, M. Pfeiffer, E.-M. Restayn, K. Ross, J. Schneider, S. Hiermaier. (2018). Validated 3D spatial stepwise quantitative hazard, risk and resilience analysis and management of explosive events in urban areas. *Eur J Secur Res.* doi:10.1007/s41125-018-0035-y.

2 Resilience of Civil Infrastructure in a Life-Cycle Context

You Dong and Dan M. Frangopol

CONTENTS

2.1 Introduction 43
2.2 Resilience Quantification..44
2.3 Resilience-Informed Assessment Incorporating Sustainability 46
2.4 Conclusions...46
References..47

2.1 INTRODUCTION

The performance of a civil infrastructure degrades due to a variety of mechanisms (such as chemical and physical attacks, corrosion, and fatigue) and natural and man-made hazards (such as earthquakes, tsunamis, hurricanes, fires, and blasts). Consequently, improving the condition and performance of infrastructure systems is a key concern. The 1994 Northridge (California) and 1995 Kobe earthquakes showed that huge economic losses are not only limited to the immediate aftermath of an earthquake but also to the long-term recovery phase. The social and economic losses can be significantly greater than the repair costs of structural components or systems and this may cause substantial harm to the community (Dong et al. 2013; Bocchini et al. 2014; Dong and Frangopol 2015). Recent events like Hurricane Katrina and Super Storm Sandy also demostrated the significant indirect impacts and economic losses, which are linked to the recovery process after a hazard event. Therefore, it is crucial to implement rational management strategies that improve performance and enhance the resilience of civil infrastructure systems to acceptable levels throughout their life-cycles.

Resilience is a term widely used in different disciplines to account for the adaptability and efficient recovery of various systems or processes. In the context of the built environment, Bruneau et al. (2003) provided a definition focusing on how social systems adapt and recover after hazard events. Presidential Policy Directive 8 (PPD-8 2011) documented a three-dimensional view of the resilience of the built environment, which it defined as a system's ability to reduce impacts, recovery time, and future vulnerabilities. This definition of resilience is prevalent in hazard-related studies, which aims to reduce direct physical damage and recovery time. Indirect losses include the additional economic losses resulting from the non-functioning of the civil infrastructure and are a function of downtime, which is affected by the damage intensity of structural and non-structural components. Many studies have been conducted on the resilience quantification of different civil infrastructure systems, e.g., water (Herrera et al. 2016), electricity, and transportation networks (Bocchini and Frangopol 2012; Sun et al. 2018), and building and building communities (Burton et al. 2015; Dong and Frangopol 2016a,b; Anelli et al. 2018). Considering the effects of uncertainty, it is crucial for the quantification of resilience at a holistic level to be processed through a probabilistic framework. Several deterministic and only a few probabilistic studies have been reported in the literature as analyzing the resilience of individual bridges and bridge networks (Decò et al. 2013; Dong and Frangopol 2017; Zheng et al. 2018). In order to predict the performance of structural systems during their life-cycle, deterioration mechanisms affecting the investigated systems (e.g., corrosion) must be carefully considered and incorporated within the resilience assessment process, taking into account the abrupt reduction of the functionality of the structures (Frangopol 2011; Frangopol et al. 2017). Thus, it is necessary to consider the resilience of a civil infrastructure against hazards in a life-cycle context.

This chapter presents a brief overview of the life-cycle resilience of civil infrastructures while accounting for uncertainty. Quantifying the life-cycle performance and resilience of a civil infrastructure at component and network levels is also discussed. This framework can serve as a useful tool for risk mitigation. The approach presented can provide optimal intervention strategies for the decision-maker. This framework will allow for resilience-informed decisions regarding the design and maintenance of civil infrastructure systems.

2.2 RESILIENCE QUANTIFICATION

Within the last few decades, the occurrence of disruptive, low-probability, high-consequences extreme events across the globe has shifted the focus of scientific communities and decision-makers to develop approaches which can improve the resilience of infrastructures against these increasingly occurring disasters. It is of great importance to incorporate resilience within structural design and assessment processes to help protect and strengthen the residing communities. Resilience is represented by its functionality and is associated with four attributes: *Robustness*: Ability to withstand an extreme event without complete failure; *Rapidity*: Ability to recover from an extreme event efficiently and effectively; *Redundancy*: Reserve or substitutive structural components or systems; and *Resourcefulness*: Efficiency in identifying problems, prioritizing solutions, and mobilizing (Bruneau et al. 2003). A resilient structure not only withstands an extreme event but also recovers efficiently and in a timely fashion. The efficiency of the recovery process is a function of the technical and organizational skills of the community, as well as social and economic conditions. Mathematically, resilience can be evaluated by integrating the functionality function over time (Cimellaro et al. 2010; Frangopol and Bocchini 2011; Bocchini et al. 2014). The three states associated with functionality can be defined as follows:

(1) Reliability state: Pre-event functionality state where a structure is considered to have baseline functionality (i.e., the civil infrastructure is functional or in an original state before the occurrence of a hazard event);

(2) Recovery state: Post-event functionality state where the structure is considered to have a loss of functionality depending upon the robustness, and time-variant functionality regain as a result of repair efforts. The repair efforts are an attribute of the resourcefulness and redundancy of the system; and
(3) Recovered state: Structure functionality after the recovery efforts (e.g., building regains functionality).

In the reliability state, a civil infrastructure (e.g., a building) has baseline functionality. The infrastructure changes from baseline functionality to residual functionality as a result of physical damage and loss of service. After a hazard event, the state changes from a reliability state to a recovery state, which includes the delay time (i.e., the time required for inspection and contractor mobilization, among others) and time-variant functionality improvement. Subsequently, repair actions are performed and the infrastructure regains its functionality and thereby reaching a recovered state.

Damage to electric power, telecommunications, transportation, and water networks due to hazard events can cause enormous social disruption (Kamissoko et al. 2018; Masoomi and van de Lindt 2018; Yang and Frangopol 2019a). Therefore, it is of vital importance to investigate the performance of these interdependent networks when subjected to hazard effects, considering that the interdependencies are on such a large scale. In order to understand the performance of these essential networks (e.g., power, communication, transportation, and water systems), their properties in terms of global connectivity, local clustering, and overall shape should be evaluated taking into account the failure modes. Then, methods and metrics to assess the performance of infrastructure networks and the interdependencies among different networks should be developed. This will contribute to the improvement of the performance-based design and management methods of interdependent infrastructure systems at the community level, considering the interdependencies between these infrastructure systems (Duenas-Osorio et al. 2007). For instance, some strong earthquakes can destroy infrastructure systems and cause injuries and/or fatalities. Therefore, it is important to investigate the seismic performance of the interdependency of the healthcare–bridge network systems to guarantee immediate medical treatment after an earthquake. The assessment of healthcare–bridge network system performance depends on the seismic vulnerability of a hospital and the bridges located in the surrounding bridge network, in addition to the ground motion intensity (Dong and Frangopol 2016b). It is important to account for the effects of the damage conditions associated with highway bridge networks on healthcare system performance.

During their life-cycle, civil infrastructure systems can be subjected to multiple hazards. Thus, during the resilience assessment procedure, it is necessary to consider performance under multiple hazards. For example, the flood-induced scour can reduce the lateral support of a bridge at its foundations. This has a major effect on the seismic bridge vulnerability. In addition, due to the effects of global warming and climate change, the frequency, intensity, and magnitude of the hazards are increasing. Since these trends are projected to continue in the near future, it is crucial to determine the effects of climate change on the resilience of deteriorating infrastructure systems (Mondoro et al. 2018). For instance, bridges located in coastal regions are the most vulnerable to adverse climate change effects. Rising sea levels combined with potentially more intense storm events and regional subsidence pose great threats to deteriorating coastal infrastructure systems and resilience assessment is becoming more and more important.

2.3 RESILIENCE-INFORMED ASSESSMENT INCORPORATING SUSTAINABILITY

Sustainability considers the impact of a structure related to its environmental, economic, and social aspects; while resilience measures the impact of an extreme event and the time it takes to recover in order to reduce direct and indirect socio-economic consequences (Bocchini et al. 2014). It is important to develop an integrated framework incorporating both sustainability and resilience (Yang and Frangopol 2019b). The probabilistic resilience can be quantified by measuring the residual functionality of a civil infrastructure and its recovery to a pre-event functionality state. The residual functionality is mapped against different limit states, with baseline functionality at "no damage" and zero functionality at "total failure." Intermediate functionalities are assigned to different damage states to account for uncertainties. Recovery functions can be linear, trigonometric, exponential, and depend on the community's resourcefulness. Recovery functions are used to track time-variant functionality improvements and, after the downtime, to achieve full functionality. The time-variant functionality over the investigated period can thus be determined, and the resilience can be assessed by integrating the time-variant functionality.

The performance-based engineering (PBE) methodology used for sustainability quantification is also an important issue. Loss analysis is performed to determine economic losses and downtime. Environmental analysis is additionally performed to determine sustainability impact (Wang et al. 2018, 2019). The environmental model is used to determine the environmental consequences of collapse and non-collapse conditions. Limit state fragilities are utilized along with the hazard model to determine the probability of failure of the civil infrastructure systems, and its related consequences (e.g., equivalent carbon emissions) are evaluated by quantifying materials used in the construction of a building. The environmental model accounts for the measurement of equivalent carbon emissions produced due to the repair and/or reconstruction of a building. Hence, seismic risk, sustainability, and resilience can be computed in an integrated manner.

2.4 CONCLUSIONS

This chapter presents a brief overview of the integration of risk, sustainability, and resilience measurements into the performance assessment of infrastructure systems under hazard effects. The chapter provides a probabilistic framework for computing seismic sustainability and resilience using a PBE methodology. The uncertainties associated with structural performance and consequence functions are incorporated. Distributed repair loss, equivalent carbon emissions, and downtime can be computed. The residual functionalities are determined probabilistically, and resilience is quantified for the investigated time period. The presented framework supports the sustainable development of infrastructure systems and provides optimal intervention strategies for the decision-maker. This framework will ultimately allow for resilience-informed decision-making.

REFERENCES

Anelli, A., Santa-Cruz, S., Vona, M., Tarque, N., and Laterza, M. (2018). A proactive and resilient seismic risk mitigation strategy for existing school buildings. *Structure and Infrastructure Engineering*, 1–15.

Bocchini, P., and Frangopol, D.M. (2012). Optimal resilience- and cost-based post-disaster intervention prioritization for bridges along a highway segment. *Journal of Bridge Engineering*, 17(1), 117–129.

Bocchini, P., Frangopol, D.M., Ummenhofer, T., and Zinke, T. (2014). Resilience and sustainability of the civil infrastructure: Towards a unified approach. *Journal of Infrastructure Systems*, 20, 0414004.

Bruneau, M., Chang, S.E., Eguchi, R.T., Lee, G.C., O'Rourke, T.D., Reinhorn, A.M., ... Von Winterfeldt, D. (2003). A framework to quantitatively assess and enhance the seismic resilience of communities. *Earthquake Spectra*, 19(4), 733–752.

Burton, H.V., Deierlein, G., Lallemant, D., and Lin, T. (2015). Framework for incorporating probabilistic building performance in the assessment of community seismic resilience. *Journal of Structural Engineering*, 142(8), C4015007.

Cimellaro, G.P., Reinhorn, A.M., and Bruneau, M. (2010). Framework for analytical quantification of disaster resilience. *Engineering Structures*, 32(11), 3639–3649.

Decò, A., Bocchini, P., and Frangopol, D.M. (2013). A probabilistic approach for the prediction of seismic resilience of bridges. *Earthquake Engineering and Structural Dynamics*, 42(10), 1469–1487.

Dong, Y., and Frangopol, D.M. (2015). Risk and resilience assessment of bridges under mainshock and aftershocks incorporating uncertainties. *Engineering Structures*, 83, 198–208.

Dong, Y., and Frangopol, D.M. (2016a). Probabilistic time-dependent multihazard life-cycle assessment and resilience of bridges considering climate change. *Journal of Performance of Constructed Facilities*, 30(5), 04016034, 1–12.

Dong, Y., and Frangopol, D.M. (2016b). Probabilistic assessment of an interdependent healthcare–bridge network system under seismic hazard. *Structure and Infrastructure Engineering*, 13(1), 160–170.

Dong, Y., and Frangopol, D.M. (2017). Probabilistic life-cycle cost-benefit analysis of portfolios of buildings under flood hazard. *Engineering Structures*, 142, 290–299.

Dong, Y., Frangopol, D.M., and Saydam, D. (2013). Time-variant sustainability assessment of seismically vulnerable bridges subjected to multiple hazards. *Earthquake Engineering & Structural Dynamics*, 42(10), 1451–1467.

Duenas-Osorio, L., Craig, J.I., and Goodno, B. (2007). Seismic response of critical interdependent networks. *Earthquake Engineering & Structural Dynamics*, 36, 285–306.

Frangopol, D.M. (2011). Life-cycle performance, management, and optimization of structural systems under uncertainty: Accomplishments and challenges. *Structure and Infrastructure Engineering*, 7(6), 389–413.

Frangopol, D.M., and Bocchini, P. (2011). Resilience as optimization criterion for the rehabilitation of bridges belonging to a transportation network subject to earthquake. *Proceedings of SEI-ASCE 2011 Structures Congress*, April 14–16, 2011, Las Vegas, NV.

Frangopol, D.M., Dong, Y., and Sabatino, S. (2017). Bridge life-cycle performance and cost: Analysis, prediction, optimization and decision making. *Structure and Infrastructure Engineering*, 13(10), 1239–1257.

Herrera, M., Abraham, E., and Stoianov, I. (2016). A graph-theoretic framework for assessing the resilience of sectorised water distribution networks. *Water Resources Management*, 30(5), 1685–1699.

Kamissoko, D., Nastov, B., Benaben, F., Chapurlat, V., Bony-Dandrieux, A., Tixier, J., … Daclin, N. (2018). Continuous and multidimensional assessment of resilience based on functionality analysis for interconnected systems. *Structure and Infrastructure Engineering*, 1–16.

Masoomi, H., and van de Lindt, J.W. (2018). Restoration and functionality assessment of a community subjected to tornado hazard. *Structure and Infrastructure Engineering*, 14(3), 275–291.

Mondoro, A., Frangopol, D.M., and Liu, L. (2018). Multi-criteria robust optimization framework for bridge adaptation under climate change. *Structural Safety*, 74, 14–23.

PPD-8. (March 30, 2011). Presidential policy directive, PPD-8 – National preparedness. The White House. Retrieved from http://www.dhs.gov/presidential-policy-directive-8-national-preparedness.

Sun, W., Bocchini, P., and Davison, B.D. (2018). Resilience metrics and measurement methods for transportation infrastructure: The state of the art. *Sustainable and Resilient Infrastructure*, 1–32.

Wang, Z., Jin, W., Dong, Y., and Frangopol, D.M. (2018). Hierarchical life-cycle design of reinforced concrete structures incorporating durability, economic efficiency and green objectives. *Engineering Structures*, 157, 119–131.

Wang, Z., Yang, D.Y., Frangopol, D.M., and Jin, W. (2019). Inclusion of environmental impacts in life-cycle cost analysis of bridge structures. *Sustainable and Resilient Infrastructure*, published online: January 9, 2019, doi:10.1080/13632469.2018.1507955.

Yang, D.Y., and Frangopol, D.M. (2019a). Life-cycle management of deteriorating civil infrastructure considering resilience to lifetime hazards: A general approach based on renewal-reward processes. *Reliability Engineering & System Safety*, 183, 197–212.

Yang, D.Y., and Frangopol, D.M. (2019b). Bridging the gap between sustainability and resilience of civil infrastructure using lifetime resilience. *Chapter 23 in Routledge Handbook of Sustainable and Resilient Infrastructure*, P. Gardoni, ed., Routledge, 419–442.

Zheng, Y., Dong, Y., and Li, Y. (2018). Resilience and life-cycle performance of smart bridges with shape memory alloy (SMA)-cable-based bearings. *Construction & Building Materials*, 158, 389–400.

3 Christchurch
Rebuilding a Resilient City?

Michel Bruneau and Gregory MacRae

CONTENTS

3.1 Introduction 49
3.2 Quantitative Findings 50
3.3 Qualitative Findings and Resilience Considerations 54
3.4 Conclusions 55
Acknowledgments 56
References 56

3.1 INTRODUCTION

On February 22, 2011, Christchurch, New Zealand, was hit by an earthquake of 6.3 magnitude with a hypocenter depth of 5 kilometers and a horizontal distance of less than 10 kilometers from the city's Central Business District (CBD). It was the strongest of a string of earthquakes that has struck the Canterbury area since 2010. The effects of these earthquakes have been extensively documented (NZSEE 2011). Access to Christchurch's Central Business District was severely restricted for months (years in some parts); since then, many of the buildings in the CBD have been demolished, and reconstruction has started. Much of the rebuilding of multistory buildings is taking place at the heart of the city (Christchurch City Council 2011); where new buildings were predominantly built of reinforced concrete prior to the earthquakes, the "new Christchurch" that is emerging is a city with a variety of structural forms. The structural systems used are diverse, ranging from traditional systems to innovative systems. This is a dramatic departure from past practice.

To quantify the extent of the shift in construction practice taking place there, and, more importantly, to identify some of the drivers that have influenced the decisions about structural material and specific structural systems, the authors have conducted a series of interviews with the structural designers of more than 60% of the post-earthquake buildings constructed to date in Christchurch's CBD, as well as with other stakeholders. Results presented in Bruneau and MacRae (2017) show that the drivers are diverse and include costs, construction speed, perceptions of damage and of structural performance, tenants requirements, engineering culture, and other factors. These are explained through the narratives obtained from the interviews.

A brief summary of some aspects of this work and key quantitative findings of this study are presented here, along with a summary of key points that have an impact on the resilience of a new post-earthquake city in a region that benefits from state-of-the-art earthquake engineering expertise and capabilities. The complete findings from this study are presented in the 170-page report "Reconstructing Christchurch: A Seismic Shift in Building Structural System," which can be downloaded from http://resources.quakecentre.co.nz/reconstructing-christchurch/.

3.2 QUANTITATIVE FINDINGS

The complete methodology used to collect the data is presented in Bruneau and MacRae (2017). In essence, the most important step of the methodology was a series of interviews conducted with structural engineering consultants. The consultants interviewed were selected based on the number of buildings their firm had designed that were constructed or being constructed as part of the Christchurch recovery. The final ten consultants selected for interviews were responsible for over 65% of the multistory buildings, and a slightly greater percentage of the total floor area, of the new buildings being constructed in the Christchurch CBD and Addington areas.

Quantitative results obtained following the methodology are presented in this section. Note that where information is presented as a function of year of consent, it must be recognized that results for 2017 are only for the first three months of the year (as data was collected, and last interviews were conducted, in March 2017). Overall, the data was obtained for a total of 74 buildings, collectively adding to a total of 482,317 square meters of floor space.

The types of lateral-load resisting systems that were included as part of this sample include: buckling restrained braces (BRB), concentrically braced frames (CBF), eccentrically braced frames (EBF), eccentrically braced frames with replaceable links (EBR), steel moment resisting frames (MRF), steel moment resisting frames with friction connections (MFF), steel moment resisting frames with reduced beam sections/"Dogbone" (MFD), reinforced concrete walls (RCW), reinforced concrete moment resisting frames (RCF), rocking frame steel (RFS), rocking frame concrete precast walls (RFC), laminated veneer lumber (LVL), base isolated buildings (B), buildings with viscous dampers (D), and building with multiple structural systems along height or in a given horizontal direction, called Hybrid (H).

Figures 3.1 to 3.5 provide some examples for a number of the above steel structural systems. More specifically, Figure 3.1 shows a building with moment resisting frames in orthogonal directions, and reduced beam sections with each beam welded to an end-plate, itself bolted to a box column. Figure 3.2 illustrates beam-to-column connections and column base connections in a building with a steel moment resisting frame and friction connections. Figure 3.3 illustrates an eccentrically braced frame with replaceable bolted links – although eccentrically braced frames were used in Christchurch prior to the 2011 earthquake, such frames with replaceable links are a post-earthquake addition to the Christchurch buildings inventory. Buildings with Buckling Restrained Braced Frames, such in the example shown in Figure 3.4, are also such a post-earthquake addition. Figure 3.5 illustrates an innovative structural system consisting of a pair of rocking braced frames linked together with criss-crossing buckling restrained braces serving as energy dissipators during the development of the rocking mechanism.

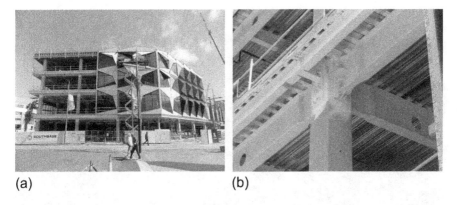

FIGURE 3.1 MRF at the crossings, 71 Lichfield Street. (a) Global view of a space moment frame. (b) Close-up of a RBS connection with bolted end-plate to moment resisting connection to square steel section.

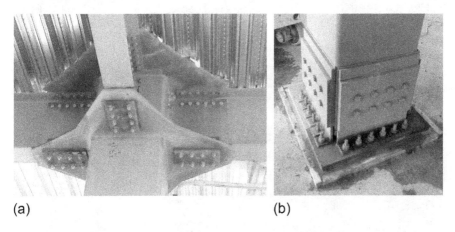

FIGURE 3.2 Details of the friction connections on the Terrace Project at the Oxford Terrace. (a) Completed bi-directional moment connection. (b) Base connection detail.

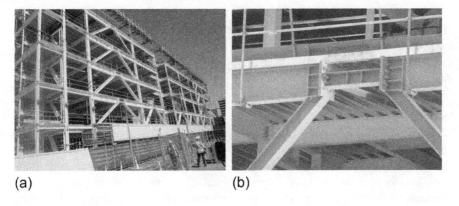

FIGURE 3.3 EBF with replaceable links, 120 Hereford Street. (a) Global view of an inverted-V braced frame. (b) Close-up of a link in an inverted-V braced frame.

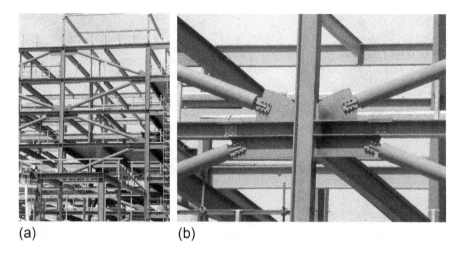

FIGURE 3.4 BRB frame, building on the northeast corner of Lichfield and Colombo Streets. (a) Global view. (b) Connection to a column at the mid-bay of a braced frame.

FIGURE 3.5 Rocking frame system implemented in the Forte Health building. (a) Coupled rocking frames. (b) Energy-dissipating couplers between rocking frames.

As part of the reconstruction, Bruneau and MacRae (2017) showed that steel, reinforced concrete, and timber lateral-load resisting systems have been used in buildings respectively totaling 377,929 square meters, 98,572 square meters, and 5,816 square meters, for a total of 482,317 square meters of floor space. This corresponds to 78.4%, 20.4%, and 1.2% of the total floor area for the three materials respectively. If the gravity systems were to be included in the above numbers, the total floor area supported by structural steel would be further increased. This is because steel gravity framing (columns, beams, floors) has been used in about 75% of the buildings with reinforced concrete walls as their lateral-load resisting system. This results in approximately 95% of the total supported floor areas in new buildings relying on steel framing.

Christchurch

To quantify the number of new buildings with different types of lateral-load resisting structural systems, the data has been broken down into the following categories for the lateral-load resisting systems mentioned previously:

- BRB: 11 total
- CBF: 3 total
- EBF: 2 total
- EBR: 4 total
- MRF: 9.5 total
- MFF: 1 total
- MFD: 4.5 total
- RCW: 32.5 total
- RCF: 0.5 total
- RFS: 1.5 total
- RFC: 0.5 total
- LVL: 2.5 total
- B: 11 total
- D: 2 total
- H: 7 total

Details of how the buildings are counted are presented in Bruneau and MacRae (2017). Information is presented in Figure 3.6 in terms of floor areas (in square meters) per structural system. Note that cumulative results are shown as grouped together for the most popular structural systems.

The results show that the following lateral-load resistance systems have been used for buildings totaling the following floor areas:

- BRB: 111,000 square meters (23%)
- CBF: 38,500 square meters (8%)
- EBF+EBR: 27,500 square meters (6%)
- MRF+MFF+MFD: 202,000 square meters (42%)
- RCW: 80,400 square meters (17%)
- RFS+RFC: 15,000 square meters (3%)

Interestingly, the 11 base isolated buildings (15% of the total number of buildings) alone provide a total of 190,000 square meters, equivalent to 40% of the total floor

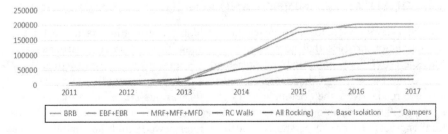

FIGURE 3.6 Growth over time in total floor area of new buildings with various types of lateral-load resisting systems (regrouped as shown).

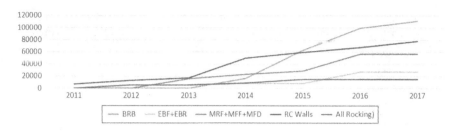

FIGURE 3.7 Growth over time in total floor area of new non-base-isolated buildings with various types of lateral-load resisting systems (regrouped as shown).

area of the buildings considered in this study. This indicates that the base isolated buildings have generally been large buildings. Indeed, the two largest base isolated buildings alone, built specifically for public sector tenants, together add-up to more 102,000 square meters (21% of the total floor area of the buildings considered here). Considering the three largest instead add-up to 129,000 square meters (and 27% of the total floor area). Also, note the strong correlation between the floor areas for base isolated buildings and steel moment resisting frames.

To better understand trends in design, Figure 3.7 shows the same results as Figure 3.6, but for all structures that have not been base isolated, as it is interesting to identify which structural systems have been more dominantly used when buildings have not been base isolated. Figure 3.7 shows that the contribution of lateral-load resistance systems to the total non-base-isolated reconstruction floor area is:

- BRB: 111,000 square meters (38%)
- CBF: 0 square meters (0%)
- EBF+EBR: 27,500 square meters (9.5%)
- MRF+MFF+MDF: 57,000 square meters (20%)
- RCW: 78,000 square meters (27%)

As such, with respect to new non-base-isolated buildings, concrete lateral-load resisting systems have been used for 27% of the floor area, and steel for 68% of the floor area, with all the other systems (i.e., hybrid structures, timber structures, rocking frames, etc.) accounting for 5%.

3.3 QUALITATIVE FINDINGS AND RESILIENCE CONSIDERATIONS

The interviews conducted also provided a valuable overarching narrative for the reconstruction process that goes beyond the quantification process. Bruneau and MacRae (2017) have compiled and presented the opinions expressed by those they met. Using the words of those interviewed (but ordering the modified sentences to improve readability), the document highlights the various factors identified as having a significant impact on the decision-making process of owners/tenants and structural engineers (from the perspective of those interviewed), and the conditions that were necessary for those factors to drive (or not), in some instances, the choice of specific structural systems. The narrative shows that some of the opinions presented

Christchurch

are contradictory to other opinions expressed, illustrating the diversity of opinions among those interviewed.

This critical part of the report (i.e., 75 of the total 170 pages) cannot be summarized without losing the critical perspective of this breath of opinions, the reasons that sustain it, and important nuances that impact decisions from case to case. However, on the basis of the above findings and interviews reported in Bruneau and MacRae (2017), the following key points can be drawn that have an impact on the resilience of a new post-earthquake city in a region that benefits from state-of-the-art earthquake engineering expertise and capabilities:

- It is becoming a more widely held belief that preventing loss of life as a seismic performance objective is simply not sufficient for a good modern structure
- Structural engineers' professional opinions impact the adoption of low-damage systems
- Tenant expectations strongly impact the choice of structural systems for individual buildings
- Additional increases in seismic performance, if desired for all buildings, would need to come from government regulation
- Context affects final decision outcome
- The reconstruction experience has paralleled an increase in stakeholder knowledge

3.4 CONCLUSIONS

With the rebuilding of Christchurch, which has taken place since 2011, the number of buildings with steel, concrete, and timber lateral force resisting systems have been in the ratio of approximately 10:10:1. However, the floor area ratios of the same buildings with steel, concrete, and timber lateral force resisting systems are about 79:20:1, because the steel systems tend to have been used in larger structures. Furthermore, for the above concrete buildings, the internal gravity frames have been found to be of structural steel three-quarters of the time.

Concrete structures in the rebuild were nearly all structural wall systems. Steel buildings have been constructed using a variety of lateral-load resisting systems. The most frequently used systems in descending order are: buckling restrained brace frames; traditional moment resisting frames; MRFs with reduced beam sections; eccentrically braced frames with replaceable links; concentrically braced frames, traditional EBFs, and rocking steel frame systems; and MRF friction frames. Most new base isolated buildings support either steel moment resisting frames or concentrically braced frames. When considering only non-base-isolated buildings, buckling restrained braces have been used in buildings adding up to nearly 40% of the total new constructed floor area.

Towards the objective of achieving resilient communities, it can be drawn from the Christchurch reconstruction experience so far that preventing loss of life is less frequently an acceptable seismic performance objective for modern buildings. The professional opinions of structural engineers have driven the adoption of low-damage systems, but tenant expectations have a significant direct or indirect impact on the choice of structural systems for individual buildings. Yet, while the reconstruction experience has paralleled an increase in stakeholders' knowledge, government regulations would still be required if the objective was to achieve an across-the-board increase in seismic performance for all buildings in a community (something yet to occur at this time).

ACKNOWLEDGMENTS

This work has been made possible by the contributions of many consultants, steel fabricators, contractors, and other individuals (listed in Bruneau and MacRae 2017) who met with the authors and were willing to share their experiences of the Christchurch reconstruction process. Their generous contributions are sincerely appreciated. The authors also thank Christchurch City Council for providing information on building consents from the city database, and Steel Construction New Zealand for kindly sharing information from their own database. The support of the Quake Centre based at the University of Canterbury in making this project possible is also sincerely appreciated.

REFERENCES

Bruneau, M., MacRae, G.A., 2017. "Reconstructing Christchurch: A Seismic Shift in Building Structural Systems," Quake Center Report, University of Canterbury, Christchurch, New Zealand.

Christchurch City Council, 2011. "Draft Central City Recovery Plan, For Ministerial Approval, December 2011," Christchurch City Council, Christchurch, New Zealand, 161p.

NZSEE, 2011. "2011 Christchurch Earthquake Special Issue," *Bulletin of the New Zealand Society for Earthquake Engineering*, Vol. 44, No. 4, pp. 181–430.

4 Resilient Bridge Decks Based on ISIS Winnipeg Principles

Aftab Mufti

CONTENTS

4.1 ISIS Winnipeg Principles 57
4.2 Application of ISIS Winnipeg Principles in the Field 57
    4.2.1 First-Generation Corrosion Free Bridge Decks 57
    4.2.2 Salmon River Highway Bridge, Nova Scotia.......................................58
4.3 Second-Generation Steel-Free Bridge Decks Based on Winnipeg Principles..... 60
    4.3.1 Red River Bridge North Perimeter Highway, Winnipeg, Manitoba.... 60
4.4 Fatigue Studies of Second-Generation Corrosion-Free Bridge Decks 62
    4.4.1 Fatigue Testing..63
    4.4.2 Performance of Bridge Decks...63
    4.4.3 Studies of Concrete Reinforced with GFRP Specimens from Field Demonstration Projects 63
4.5 Conclusions and Recommendations 65
Acknowledgments...66
References..66

4.1 ISIS WINNIPEG PRINCIPLES

The search for innovation in civil engineering has led to the creation of the Winnipeg Principles for the design of bridge decks using an arching action. The Winnipeg Principles, which were developed at the International Workshop on Innovative Bridge Deck Technologies held in Winnipeg, Manitoba, in April 2005, and adopted as guidelines for concrete bridge decks by practicing engineers, held that concrete bridge decks should preferably be designed in accordance with the inherent arching action present, whereby the top reinforcement is no longer required for strength, and that new code provisions should permit the use of FRPs as reinforcing materials for civil engineering structures.

4.2 APPLICATION OF ISIS WINNIPEG PRINCIPLES IN THE FIELD

4.2.1 First-Generation Corrosion Free Bridge Decks

In the quest for lighter, stronger, and corrosion-resistant structures, the replacement of ferrous materials by high-strength fibrous ones is being actively pursued in several countries around the world, both with respect to the design of new structures as well as for the rehabilitation and strengthening of existing ones. In the design of new highway bridges in Canada, active research has been focused on a number of specialty areas, including the replacement of steel reinforcing bars in concrete deck slabs by randomly distributed low-modulus fibers, and the replacement of steel prestressing cables in concrete components by tendons comprised of super-strong fibers. Research is also being conducted to repair and strengthen existing structures with the use of FRPs.

FRPs have perceived disadvantages compared to steel. These are the related to the ductility and low thermal compatibility between FRP reinforcement and concrete. The majority of our construction projects in Canada are in non-seismic zones. Ductility is an important characteristic of steel as it allows large deformations and the dissipation of energy. Concrete structures reinforced with FRPs at ultimate loads give large deformations. Therefore, reinforced concrete structures, whether reinforced with steel bars or FRPs, give the same order of deformability. Research to show that concrete structures with FRPs, if properly designed, can dissipate energy is in progress. The design of the proper concrete cover eliminates low thermal compatibility between FRP reinforcement and concrete. It should be noted that glass fiber reinforced polymer (GFRP) material has a modulus of elasticity comparable to concrete. Therefore, concrete does not feel any intrusion and performs well in resisting fatigue under dynamic loading.

These concepts have been implemented to develop corrosion-free bridge decks. Several such bridge decks have been constructed in Canada and one in Iowa, USA. One of these bridge decks is described in the following section.

4.2.2 SALMON RIVER HIGHWAY BRIDGE, NOVA SCOTIA

The first-generation steel-free deck slab in Canada was cast on the Salmon River Bridge, part of the Trans-Canada 104 Highway in Nova Scotia [1]. Construction of the bridge, which consists of two 31-meter spans, includes a steel-free deck over one span and a conventional steel-reinforced deck over the other. Internal arching in the slabs helps transfer the loads to the girders. Although the cost of the steel-free side was 6% more than the steel-reinforced side, the overall design tends to be less expensive than conventional decks. This is because steel-free decks do not suffer from corrosion, so traditional maintenance costs are greatly reduced. This concept has won six national and international awards, including the prestigious NOVA award from the Construction Innovation Forum (CIF) of the United States.

The deck contains no rebar. Instead, longitudinal beams or girders support it. The load is transferred from the deck to the supporting girders in the same way that an arch transfers loads to supporting columns. Although steel straps are applied to tie the girders together, because they are not embedded in the concrete, they can be easily monitored and inexpensively replaced.

The structural health monitoring of the steel-free bridge deck was conducted by installing sensors (Figure 4.1). SHM indicates that the load sharing of the Salmon River Highway Bridge is similar to conventional decks (Figure 4.2).

With no steel inside the concrete (Figure 4.3), no unnecessary weight is added, meaning thinner deck designs. The steel straps are welded to the top flanges of the

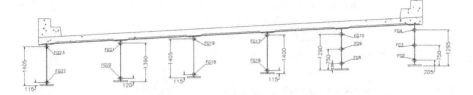

FIGURE 4.1 Sensor locations.

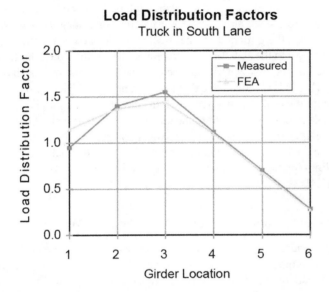

FIGURE 4.2 Load sharing of the Salmon River Highway Bridge.

FIGURE 4.3 Casting of the steel-free deck.

girders, thereby resisting any lateral movement. The Salmon River steel-free bridge deck has withstood a number of Canadian winters, and it appears to be defying the conventional approach to building steel-reinforced bridge decks. There are 11 steel-free bridge decks now in service across Canada.

4.3 SECOND-GENERATION STEEL-FREE BRIDGE DECKS BASED ON WINNIPEG PRINCIPLES

The second-generation steel-free deck slab exhibits the same behavior as the first-generation steel-free deck slab, with the exception of a longitudinal crack at the mid-point between the girders. External steel straps located below the deck provide structural integrity to the slab. In order to reduce the width of the longitudinal crack that developed on the first-generation steel-free decks, researchers at the University of Manitoba (U of M) [2] concluded that a bottom mat of GFRP reinforcement with a reinforcement ratio of 0.25% was required. In addition, recent fatigue tests have also been undertaken at the U of M to replicate actual service life conditions for a deck slab. These tests confirmed that a steel-free deck slab reinforced with a crack control grid of nominal GFRP reinforcement exhibits a maximum crack width of 0.34 millimeters, a limit implicitly acceptable to the Canadian Highway Bridge Design Code, (CHBDC) [3,4].

4.3.1 Red River Bridge North Perimeter Highway, Winnipeg, Manitoba

This ten-span bridge is 347 meters long and consists of steel plate girders, spaced at 1.8 meters and a composite, cast-in-place, steel-reinforced concrete deck. It is located on the north half of the Perimeter Highway that encircles the City of Winnipeg. Because the Perimeter Highway forms part of the Trans-Canada Highway system, this bridge is subjected to significant daily traffic loads with approximately 20% being truck traffic.

The one span utilizing the second-generation steel-free deck technology was designed and cast using a concrete deck slab thickness of 200 meters. GFRP reinforcement was used for both the top and bottom mats in the internal deck panels. The top and bottom transverse and longitudinal reinforcement were comprised of #3 bars spaced at 200 and 600 millimeters, respectively (Figure 4.4). CFRP reinforcement was used as the main reinforcement in negative moment regions for both the

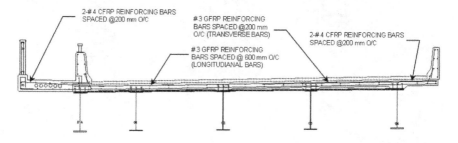

FIGURE 4.4 Top and bottom transverse and longitudinal reinforcement.

vehicular and pedestrian cantilevers. This transverse reinforcing consisted of 2– #4 bars spaced at 200 millimeters.

Transverse confinement of the deck slab was provided by steel straps, measuring 50 millimeters in width by 30 millimeters in depth, that were tack welded to the top flanges of the steel plate girders at a spacing of 1.2 meters. To ensure that the steel straps would perform integrally with the deck slab, steel Nelson studs were added to the straps in the portion that passed over the girders.

For the Red River Bridge, the integrated SHM, or civionics system, was designed and installed to monitor the components of the steel-free deck slab and provide data on the stresses in the GFRP reinforcement and the transverse steel straps. Stresses in the steel plate girders and the CFRP reinforcement in the negative moment regions for the cantilever sections are also monitored. The system is comprised of a combination of various types of sensors, namely, conventional electric strain gauges, fiber optic Bragg sensors, accelerometers, and thermocouples. A portion of the civionics system used for the Red River Bridge is shown in Figure 4.5. Figures 4.6a and b show some of the civionics instrumentation installed by Vector Construction prior to casting of the bridge deck.

Each sensor is connected to an on-site data acquisition system capable of storing the data. In addition, a video camera and weigh-in-motion device are also being installed to gather additional information regarding the frequency, configuration, and axle loading of truck traffic on this specific bridge.

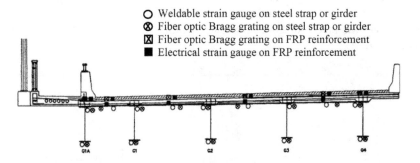

FIGURE 4.5 Civionics system for the Red River Bridge North Perimeter.

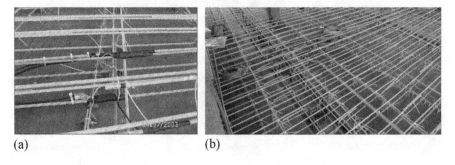

FIGURE 4.6 Installed civionics instrumentation.

One major concern of a civionics system is the enormous quantity of data that can be generated and must be stored in a short period of time. Sensors typically can take up to 100 readings per second, resulting in 8.64 million readings in a single day per sensor. The Red River Bridge contains 64 sensors, which translates into 0.5 billion readings per day. ISIS Canada, in conjunction with IDERS Incorporated, is currently developing an automated system that can be incorporated into the SHM unit. The readings will be scanned for pre-determined strain readings that will send a "red flag" notification to the design engineer. The automated system will greatly reduce the time and cost required to review the entire load history of the deck span.

4.4 FATIGUE STUDIES OF SECOND-GENERATION CORROSION-FREE BRIDGE DECKS

This section describes the fatigue behavior of cast-in-place second-generation steel-free bridge decks. Although cast monolithically, the bridge deck was divided into three segments (A, B, and C; see Figure 4.7 for the details). Segment A was reinforced according to conventional design with steel reinforcement. Segments B and C were reinforced internally with a carbon fiber reinforced polymer (CFRP) crack control grid and a GFRP crack control grid, respectively, and externally with steel straps. The hybrid CFRP/GFRP and steel strap design is called a second-generation steel-free concrete bridge deck. All three segments were designed with an almost equal ultimate capacity so that a direct comparison between the segments under fatigue loading conditions could be made. A performance comparison of all three segments for the first bridge deck under 60-ton (588 kN) cyclic loads is reported in this chapter.

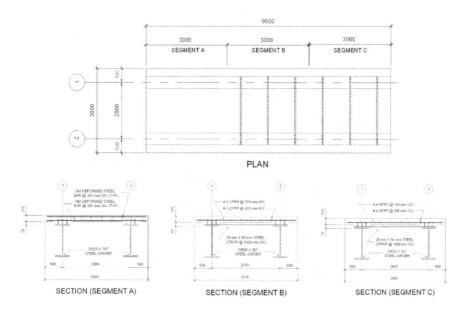

FIGURE 4.7 Bridge deck reinforcement details.

Resilient Bridge Decks Based on ISIS Winnipeg Principles

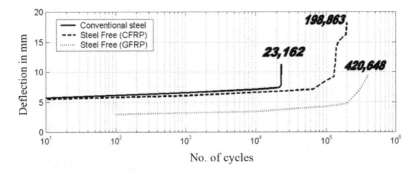

FIGURE 4.8 Plot of deflection versus number of cycles at 60 tons.

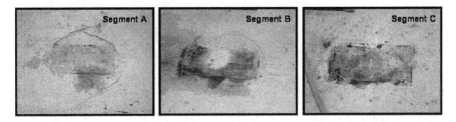

FIGURE 4.9 Punching modes of fatigue failure for segments A, B, and C.

4.4.1 Fatigue Testing

Figure 4.8 illustrates the crack width behavior for all three bridge deck segments under the 60-ton or 588-kN load level. The results show that deck Segment A fatigued approximately 20 times as fast as deck Segment C, and deck Segment B fatigued approximately twice as fast as deck Segment C. All three segments failed in fatigue and via a punching shear (Figure 4.9).

4.4.2 Performance of Bridge Decks

The performance of bridges should be measured in terms of fatigue damage and number of cycles. Fatigue damage could be quantified by either the cradle width or the deflections. It should be noted that, at failure, the ultimate deflections measured under static or dynamic load are the same, as is shown in the load deflection curve in Figure 4.10. Also, performance is judged by the number of cycles and crack width (Figure 4.11). As soon as the cycles approach a critical region, strengthening should begin.

4.4.3 Studies of Concrete Reinforced with GFRP Specimens from Field Demonstration Projects

The methods used in this study [5] to investigate the degradation of GFRP reinforced concrete are Scanning Electron Microscopy (SEM) and Energy Dispersive X-ray (EDX), Light Microscopy (LM), Differential Scanning Calorimetry (DSC), and Infrared Spectroscopy. To obtain reliable information from the results/observations

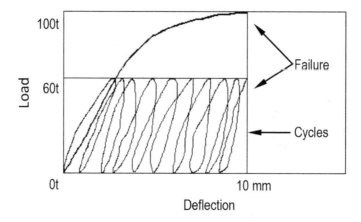

FIGURE 4.10 Load and deflection performance curves for static and dynamic loads.

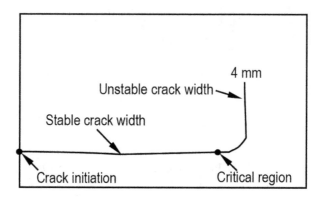

FIGURE 4.11 Crack width performance under dynamic loads.

using such methods, special attention was given to sample preparation. During specimen preparation, the glass fiber can be scratched and microcracks can be induced into the matrix and concrete; the glass fiber can be debonded; and the glass and matrix polished surfaces can be contaminated with elements from each other and with elements from the concrete. When such events take place, the interpretation of the results/observations becomes laborious.

Details on sample preparation, presentation of the results obtained on the entire set of GFRP reinforced concrete specimens removed from the five structures, and an extended discussion of the results were presented at the 3rd International Conference on Composites in Construction 2005 in Lyon, France [6] and will be further detailed in the ISIS Canada Monograph on GFRP durability [5].

The following is a brief excerpt of the main findings suggested by the results obtained to date from the analysis performed by Onofrei [7] at the University of Manitoba on the randomly selected core specimens from field demonstration projects using SEM and EDX analyses. Examples of a SEM micrograph and EDX spectra on GFRP specimens cored from the Joffre Bridge are shown in Figure 4.12a and b.

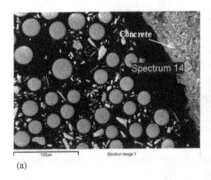 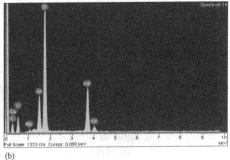

FIGURE 4.12 SEM photomicrograph (a) the cross section of core specimen from Joffre Bridge. EDX spectrum (b) shows the chemical composition of the glass fiber—Magnification factor is 500×.

Although the entire surface of each specimen was examined, attention was focused on areas close to the concrete/GFRP rod interface, where any degradation could be expected to take place.

Data from SEM analyses suggests that there were no visible signs of any degradation taking place on the reinforced bars/grids. The individual fibers are intact with no gaps between the fibers and the matrix (Figure 4.12a). There is also good contact between the concrete and the matrix.

It can be concluded that for the range of conditions of field demonstration projects included in this study there is no degradation of the GFRP reinforcement. Since the pH of the concrete pore water solution is expected to decrease further with time, it is quite probable that for all practical problems the degradation of GFRP can be considered to be insignificant.

The results obtained to date in this study do not support the results obtained in laboratory studies that report GFRP degradation in various degrees. There are several reasons for this disagreement. First, in the present study, the specimens under investigation were from real-life engineering structures; whereas in the laboratory studies that observed GFRP degradation, the conditions were very different from the in-service systems. Lab studies are usually conducted in alkaline solutions that are kept at constant high pH values and/or high temperatures. The GFRP is generally kept in contact with an infinite leachate volume or a very large effective surface area of the rod is exposed to the solution. With the exception of special projects, such conditions do not normally exist in real life. Therefore, for most practical engineering projects, it is quite probable that the degradation of GFRP reinforced concrete in most cases is not significant.

4.5 CONCLUSIONS AND RECOMMENDATIONS

As mentioned earlier, ISIS Canada intends to significantly change the design and construction of civil engineering structures. For changes in design and construction to be accepted, it is necessary that innovative structures are monitored for their health. To assist in achieving this goal, ISIS Canada is developing a new discipline, which integrates civil engineering and electronics under the combined banner of civionics.

The new discipline of civionics must be developed by civil structural engineers and Electronics Engineers to lend validity and integrity to the process. Civionics will produce engineers with the knowledge to build "smart" structures containing the SHM equipment to provide much-needed information related to the health of structures before things go wrong. SHM has shown that decks behave like arches and not like plates. Hence, decks should be designed according to the Winnipeg Principles.

Lastly, steel is too stiff (modulus is seven times) compared to GFRP (modulus is the same as concrete). Therefore, GFRP should be the preferred reinforcement for concrete when considering ductility and fatigue performance.

ACKNOWLEDGMENTS

The authors would like to acknowledge the financial support of ISIS Canada, NSERC, and NCE for the research work conducted. In particular, they wish to acknowledge the collaboration of Dr. Baidar Bakht, Dr. Gamil Tadros, Dr. John Newhook, and Dr. Amjad Memon on the development of the corrosion-free bridge deck concept, which has been incorporated into the Winnipeg Principles. In addition, the assistance of Ms. Nancy Fehr, Executive Assistant to the President of ISIS Canada in the editing and formatting of the chapter is gratefully acknowledged.

REFERENCES

1. Newhook, J.P. and Mufti, A.A., "A reinforcing steel free concrete bridge deck for the Salmon River Bridge," *Concrete International* **18**(6) (2000) 30–34.
2. Memon, A.H., "Fatigue Behaviour of Steel-Free Concrete Bridge Deck Slabs under Cyclic Loading," PhD Thesis, University of Manitoba, Winnipeg, Manitoba, Canada (2005).
3. Klowak, C., "Static and Fatigue Behaviour of Bridge Cantilever Overhangs Subjected to a Concentrated Load," MSc thesis to be submitted to the University of Manitoba (in progress) Winnipeg, Manitoba, Canada (2007).
4. Mufti, A.A., Tadros, G., and Bakht, B., "Design Calculations for Steel Free Deck for Red River Bridge," Report submitted to EarthTech Consultants, Winnipeg, Manitoba, Canada (2003).
5. Banthia, N., Benmokrane, B., and Karbhari, V., Editors, "ISIS Monograph on Durability," ISIS Canada Research Network, Winnipeg, Manitoba, Canada, in press (2007).
6. Mufti, A.A., Onofrei, M., Benmokrane, B., Banthia, N., Boulfiza, M., Newhook, J., Tadros, G., Bakht, B., and Brett, P., "Studies of Concrete Reinforced with GFRP Specimens from Field Demonstration Projects," Proceedings for the Third International Conference on Composites in Construction (CCC 2005), Lyon, France (2005).
7. Onofrei, M., "Durability of GFRP Reinforced Concrete from Field Demonstration Structures," ISIS Canada Research Network, Winnipeg, Manitoba, Canada (2005).

5 Resilience Considerations of a Historical Timber Bridge

Teruhiko Yoda and Weiwei Lin

CONTENTS

5.1 Introduction 67
5.2 Resilience of Bridge Structures 69
5.3 Evaluation of Resilience in the Kintaikyo Bridge 70
    5.3.1 Safety Countermeasures 70
    5.3.2 Serviceability Countermeasures 72
    5.3.3 Durability Countermeasures 74
    5.3.4 Damage Reduction Countermeasures 74
5.4 Concluding Remarks 77
References 77

5.1 INTRODUCTION

The first Kintaikyo Bridge was constructed in 1673 (the current Kintaikyo Bridge is the fourth). The Kintaikyo Bridge, a five-span timber bridge, is 193.3 meters long and 5 meters wide (with a 4.3-meter effective width of the road). The three central spans are arch bridges, and the other two end spans are warped girder bridges. The span of each arch bridge is 35.1 meters; each girder bridge is 34.8 meters. The girder of each span comprises first through eleventh girder members, a large ridge board and a small ridge board. The rear ends of the first through fourth girder members are inserted and bolt-clamped in the iron shoe mounted on the upper part of the substructure (Figures 5.1 and 5.2).

The fifth through eleventh girder members are longitudinally staggered so that each member protrudes by approximately one-third of its length from the girder member immediately beneath it. Girder members are sequentially installed in this way from each end of the span. A large ridge board is mounted between the ninth girder members from both ends; a small ridge board is mounted between the tenth girder members which are installed on the large rider board. A long thin wedge is inserted between each overlapping girder member, causing the front-end portion of each member to bend slightly downward, so that an arch is formed by all girder members. To prevent girder member displacement, dowels are placed on the surface of each girder member that contacts with the other girder members. The overlapping girder members are bound together using pairs of C-shaped hoop irons, called steel

FIGURE 5.1 Kintaikyo Bridge as of 1998.

FIGURE 5.2 Side view of the arch.

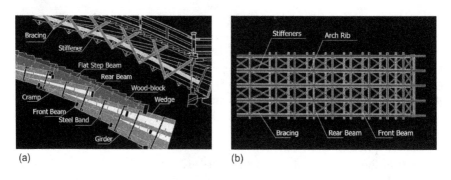

FIGURE 5.3 Structural components of the arch bridge.

bands, which are positioned on the lateral sides of the girder members. This assembly technique, unique to the Kintaikyo Bridge, is called the voussoir arch method. The original Kintaikyo Bridge constructed in 1673 was one such girder assembly structure. Ten years later, in 1683, V-shaped bracings (unique to the Kintaikyo Bridge) were installed, and auxiliary bar stiffeners were installed along each arch rib (Figure 5.3), to complete the girder assembly structure (Iwakuni City Economic Department, 2007). As the floor follows the curve of the arch, the bridge is only suitable for a pedestrian bridge.

From above, the bridge structure looks like a solid rib because it looks like a single arch rib with an elongated beam (small-size timber) overlapping. However, when approaching, the cross-section of the arch rib is composed of a superposition of three or four girder members. The superposed girder members are fastened by dowels, cramps, and steel bands. From a bridge engineering perspective, it can be called a "wooden semi-2-hinge arch bridge with large cross-section arch ribs overlapping with small-size timber struts (straight timber)."

Despite the bridge's unique five-span arch structure, which is designed to enhance resiliency, the Kintaikyo Bridge, which is primarily made of wood, is vulnerable to natural disasters.

The kinds of timbers used for this bridge, locations of their use, sizes, and other details are specified in the ancient drawing created in 1699. These specifications are followed even today. The skill of identifying timber characteristics, cultivated over many years by the Japanese wooden craft culture, enabled the use of appropriate types of timber in the right places, taking advantage of the characteristics of hard wood, soft wood, decay-resistant wood, and so on. The resulting Kintaikyo Bridge represents the essence of wooden craft culture, in addition to the beauty of the appearance with its Catenary-like curve (Figure 5.4).

In this chapter, the knowledge based on past disaster experience and the damage reduction technology to keep up with resiliency are discussed, with due consideration to the resilience of the bridge.

5.2 RESILIENCE OF BRIDGE STRUCTURES

Concepts such as redundancy and robustness play a significant role in investigating the resilience of bridge structures. Redundancy is understood as an important factor in many design areas. The idea of redundancy in bridge structures was initially raised in the early 1970s. The redundancy of bridge superstructures has been investigated

FIGURE 5.4 Form of the arch.

individually in different research institutions in different countries. The relationship between reliability and redundancy forms an integral part of bridge design and maintenance, in which reliability is understood as a reference for studying the concept of redundancy. The main reason for this is that the evaluation and determination of the reliability of bridge structures is mainly investigated at an ultimate limit state, which is the boundary between safety and failure (Nowak, 2004). Research into redundancy in a failure state shows a close resemblance to the investigations carried out in reliability studies. It follows from this that the reliability index can be used to determine an ideal model of redundancy.

Redundancy of a bridge superstructure is usually considered after its failure or collapse. The failure and collapse of a bridge superstructure can be caused by many factors, such as deterioration, excessive weight, corrosion, natural disaster, traffic accident, etc. In order to approve the economic provision and avoid huge costs of maintenance and repair, the need for redundancy in bridge structures appears. The problems to be considered are that the repair and maintenance of bridge structures are hard to provide to all the bridges homogeneously. However, research on redundancy in specific situations is difficult to extrapolate to the being relevant to all bridges. The idea that only one faulty member can cause the failure of the whole structure is not appropriate for determining redundancy because the bridge is not supported by separate elements, so the analysis of the bridge's behavior has to be performed from a more general view.

On the other hand, observation and investigation start before failure; displacement is observed and investigated for the repair and maintenance of a bridge. The idea of the concept is understood in the word *robustness*. Eurocodes documents (COST, 2011) define robustness as the "ability of a structure to withstand events like fire, explosions, impact or the consequences of human error, without being damaged to an extent disproportionate to the original cause." The idea is different but approaching state of the bridge is similar to the idea of redundancy. The concepts of redundancy and robustness have a slight difference between them. They share some of their common parts, including the use of reliability analyses in the evaluation process and the focus on structural response instead of member capacity. Part of the concept of robustness is included in redundancy but total idea has a slight conflict. Clear specifications and definitions are needed in future research and study.

5.3 EVALUATION OF RESILIENCE IN THE KINTAIKYO BRIDGE

5.3.1 Safety Countermeasures

The arch shape of the Kintaikyo Bridge changes over time after completion. The temporal change includes a portion that changes in a short period of time and a portion that changes in a long period of time, such as a creep phenomenon. Since the arch structure of the bridge is mainly made of wood, it allows the shape to change in this way, and the wood near the supports will corrode when it rains. Corrosion and weakening cause the wood to shrink, shortening the overall length of the arch and lowering the center of the arch. However, in the case of an arch structure, even if the height of the arch rib decreases, the bridge does not collapse but it does result in lowering. This yields resiliency utilizing the feature of the arch structure that the total length of the arch can be shortened without changing the distance between the two bridge piers.

Of course, this background needs to ensure that the material strength of the wood itself does not deteriorate rapidly. Therefore, in 2002 the materials were tested using the parts removed at the time of disassembly in 2001. The results of the tensile compression tests of wood and the tensile tests of metal revealed that the strength and rigidity of the wood and metal materials had not significantly deteriorated from the initial stage of construction except for corrosion, and sufficient performance was maintained. It can be understood that this led to the long-term assurance of the safety of the arch structure of the Kintaikyo Bridge.

Periodic surveys were conducted at both the Kintaikyo Bridge of Showa, which was completed in 1953, and the Kintaikyo Bridge of Heisei, which was completed in 2004, to ensure the safety of the bridges and prolong their lives; by observing aging changes, efforts are being made to detect deterioration at an early stage.

The scope of the strength survey covered three central arch bridges. It is obvious that the loaded girders at both ends of the bridge have more strength than the three arch bridges in the center, and their structure is not unique. Therefore, the strength survey was carried out only for the arch bridge from the first survey of the Kintaikyo Bridge in Showa. However, the investigation of the decay of wooden parts covers all bridges (Waseda University, 2004).

A total of nine periodic surveys were conducted on the Kintaikyo Bridge in Showa from 1953 to 1998. Since the load itself used in the periodic survey was smaller than the strength of the successive bridge, the amount of displacement was inevitably small, and even if the measurement was performed without changing the measurement method, the influence of the measurement error, etc., could be ignored. However, the safety and usability of the successive bridges can be examined by observing change over time of the maximum displacement and the frequency. The investigations of the Kintaikyo Bridge in Showa, which is an important cultural asset, proved to be very useful for protecting the bridge; by carrying out the periodic investigations, an increase of the maximum displacement was found, which led to the early detection of aging. The structural characteristics of the Kintaikyo Bridge became clearer with the continuous observation of the aging of the arch structure.

The objectives of the assessment were to estimate the present load carrying capacity, to identify any structural deficiencies in the original design, and to determine reasons for existing problems identified by the inspection. Load-carrying capacity is an important aspect affecting the safety of the bridge. Pedestrian bridges are no exception. Information regarding the ultimate strength of the bridge is required for appropriate allocation of bridge maintenance funds. Measurements of the response to static loading may be used to measure the elastic response of the bridge. However, this type of test requires significant extrapolation of the measurements, if used to predict the strength at design load level.

It indicated that the bridge is expected to be safe at least up to the design pedestrian load (600 kN), which was confirmed by the loading tests in 2001. The Kintaikyo Bridge is a wooden arch bridge unlike any other in the world. Even by modern engineering standards, the structure of this bridge is considered to be extremely advanced. Tests verified its exceptional strength (Figure 5.5). One test placed a uniform load of 60 tons

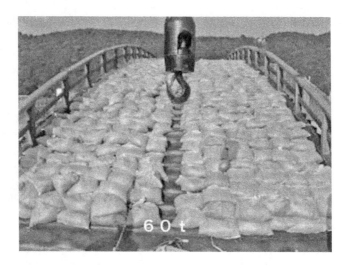

FIGURE 5.5 Uniform load of 60 tons.

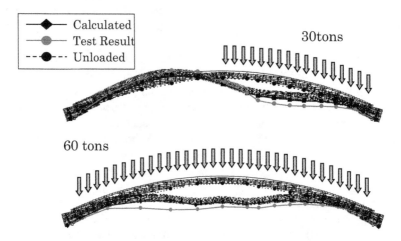

FIGURE 5.6 Comparison between tests and numerical analyses.

on the center span of the bridge before it was disassembled in 2001. Despite its then advanced age, the center of the bridge sunk only 27 millimeters under the load, as shown in Figure 5.6. This result satisfies the present standards for a pedestrian overpass.

5.3.2 Serviceability Countermeasures

Dynamic excitation of the bridge is used to identify suspected deterioration of the bridge. The use of structural modes to evaluate the integrity of the bridge was investigated. The cut-off frequency for filtering should be selected, taking into consideration the frequency response of the bridge. Dynamic tests were conducted in which three students were positioned side by side. The dynamic tests suggested that there was little dynamic amplification on the arch.

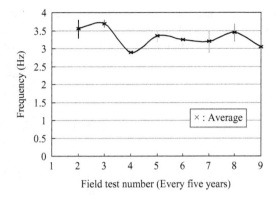

FIGURE 5.7 Lowest natural frequency in the chronological order.

FIGURE 5.8 Details of the deteriorated arch rib-to-beam joints.

The lowest natural frequency is shown in Figure 5.7. This figure shows a correlation of stiffness for evaluating the integrity of timber structures. Since the bridge of the Showa era (completed in 1953) was built more than 45 years ago, it was clear that the rib-beam joints had deteriorated, as shown in Figure 5.8.

From the point of view of bridge design, since the arch bridge is a pedestrian bridge, the limit state of the arch structure is a measure of the serviceability limit state rather than the ultimate limit state. On the pedestrian bridge, in consideration of the safety of the pedestrians, the bridge is prevented from shaking at a frequency close to 2 Hz (1.5 Hz to 2.3 Hz) when people are walking across it. Since an arch bridge is liable to swing in a horizontal direction, serviceability is limited when the natural frequency of 3 Hz approaches 2 Hz. One of the reasons for the decision to replace the natural frequency was a decrease in the natural frequency.

In the Kintaikyo Bridge, members such as saddle wood and auxiliary wood were devised to make it difficult to shake. Although it was not confirmed by the experiment, it was confirmed by the structural analysis that the bridge deck and the railing greatly affected the rigidity of the bridge.

5.3.3 Durability Countermeasures

A field inspection was performed to identify and assess deterioration. Conditions varied throughout the bridge, with most of the deterioration due to decay caused by the accumulation of dirt and moisture.

In the case of the Kintaikyo Bridge, the contrivance for long-lasting timber bridges is found in many places. Firstly, a copper plate is attached to the top surface of the girder and the wooden surface of the crossbeam, and the wooden surface is waterborne. The inclination of the bridge surface is also advantageous in the sense that it does not cause water to stagnate. As a result, water falls on both ends of the arch bridge. Even if a part of the girder end is corroded and shortened, the arch structure consists of a soft arch which is a tough arch shape, and also contributes to the improvement of durability.

In addition to the arch structure, various technologies are illuminated by the Kintaikyo Bridge. Six types of wood are used in the Kintaikyo Bridge (Figure 5.9). It is considered probable that the types of woods were changed so that the bridge lasted longer. Whether or not six types of wood were used from the outset is unknown. However, if the same wood was used, it is not possible to distinguish between rigidity, strength, and corrosion resistance. Therefore, red pine, Japanese cypress, Zelkov, and oak were used based on the idea of the right wood in the right place.

5.3.4 Damage Reduction Countermeasures

The Kintaikyo Bridge was washed away on September 14, 1950 (Figure 5.10), when the river's flow volume increased to 3,700 m3 per second at the point of the bridge. This flow volume exceeded the bridge's designed flood control level

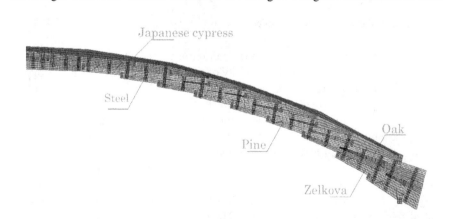

FIGURE 5.9 Girder assembly technique (Half model).

Resilience Considerations of a Historical Timber Bridge

FIGURE 5.10 Loss of one pier before collapse (1950).

of approximately 2,470 m3 per second. In 2005, Typhoon 14, which was formed between the night of September 6 and the dawn of September 7, 2005, brought heavy rainfall (maximum precipitation: 59 millimeters/hour) over the upper reach of the Nishiki River. The daily precipitation reached 472 millimeters, the heaviest rainfall in recorded history in Japan at that time. At the observing station situated 600 meters downstream from the Kintaikyo Bridge, the water level reached 7.32 meters at 1:00 a.m. on September 7, as compared to the danger level of 6.4 meters. The flow volume at that time was approximately 5,400 m3 per second, which was much greater than the level recorded when the bridge was swept away in 1950. Typhoon 14 inflicted extensive damage on the Kintaikyo Bridge (Figure 5.11). The drift from the upper reaches destroyed two piers of the first span (Figure 5.12).

FIGURE 5.11 Arches being washed away (1950).

FIGURE 5.12 Fourth pier was washed away. (News: http://www.chugoku-np.co.jp/News/index.html).

FIGURE 5.13 Loss of two piers reducing damages (2005).

This damage was very similar to the damage in 1950, as shown in Figure 5.10 (Iwakuni City Economic Department, 2007).

Despite the loss of the two piers from the first span, its superstructure remained intact (Figure 5.13). The superstructure was protected by the design of a special tenon called hozo, which had been tapered to allow members to disjoint easily, so that substructural damage would not impact the superstructure. In other words, the hozo was designed to function in a way similar to a fuse in an electric circuit. This unique damage reduction countermeasure was developed over the long history of the bridge, which experienced frequent storms and floods.

5.4 CONCLUDING REMARKS

While the present chapter only covers a fraction of the resilience considerations of the Kintaikyo Bridge, it is possible to glean some interim conclusions from the unique arch structure, which draw upon both existing knowledge on resilience and empirical technology, leading to a new era of timber bridge construction. The following conclusions can be drawn from the present study.

1. As far as present considerations are concerned, arch structures with many movable structural components under a simple stress condition such as compression are considered to be resilient. This comes from the fact that a uni-axial state guarantees stress continuity, even if there are some deteriorated parts, stress continuity is relatively assured. This is completely different from the cases of girder structures under bending stress and shear stress conditions.
2. The historical timber bridge (Kintaikyo Bridge) is considered to be a flexible arch structure in comparison with a rigid masonry arch structure, resulting in a resilient bridge. The flexible arch structure has the characteristics of redundancy simultaneously covering internal redundancy, structural redundancy, and load path redundancy.
3. The concept of resilience has different contents in different bridges. The unified and detailed concepts of resilience in bridge design should be improved or unified.

REFERENCES

A.S. Nowak (2004). System Reliability for Bridge Structures. University of Nebraska, Lincoln, NE68588-0031.

European Cooperation in Science and Technology (COST) (2011). Robustness of Structures. Action TU0601, Brussels, Belgium.

Iwakuni City Economic Department (2007). Kintaikyo: Bridge That Links Past and Future, Pamphlet for Foreigners.

Waseda University's National Research Center for Science and Technology (2004). Kintaikyo Bridge Strength Test Report (in Japanese).

6 Resiliency and Recoverability of Concrete Structures

Zhishen Wu and Mohamed F.M. Fahmy

CONTENTS

6.1 Introduction 79
6.2 Sustainability Design of Urban Structures 80
6.3 Model for Sustainability Performance-Based Design of RC Structures 81
6.4 Sustainability, Resiliency, and Recoverability.. 84
6.5 What Global Efforts are Being Made to Enhance the Resiliency of RC Structures to Move Toward Fully Sustainable Countries? 85
    6.5.1 The International Organization for Standardization (ISO) 86
6.6 What is the Situation of Current RC Structures?.. 87
6.7 The Importance of Enhancing the Recoverability of RC Structures 87
6.8 Efforts Done to Enhance Recoverability/Restorability of RC Structures 88
    6.8.1 Modern Seismic Design Codes.. 88
6.9 Application of FRP Composites to Enhance the Recoverability of RC Structures... 91
6.10 Recoverability of Existing Structures Using FRP Composites 92
    6.10.1 Limit States of Existing Structures after Retrofitting Using FRP Composites.. 96
6.11 Modern Recoverable FRP-Steel Reinforced Structures 98
    6.11.1 Innovative Steel Fiber Composite Bars (SFCB) 99
    6.11.2 FRP Bars as Longitudinal Reinforcement and Steel Reinforcement 99
    6.11.3 Bond-Based Design Controllable FRP Bars as Longitudinal Reinforcement and Steel Reinforcement 101
    6.11.4 Innovative Resilient Systems Using FRP Composites 104
6.12 Conclusions... 105
References... 106

6.1 INTRODUCTION

Human-constructed structural systems in normal operational conditions are operating under a daily load, and during long-term service they become subjected to environmental erosion and therefore will become overloaded. In addition, many structures are at risk of accidental effects, i.e., natural hazards or human disasters. All these factors cause gradual or immediate degradation of the performance of the available systems. In order to ensure that existing and modern structural systems perform the required functions and services, and retain their capacities during the life-cycle, numerous academic and field studies as well as lessons learned in times of crisis and natural hazards have produced detailed design recommendations and guidelines, taking into account several parameters: changes in the applied loads, weather impacts, and exposure to unconventional actions. At present, there is a general consensus among the academic community, professional engineers, expert construction industries, and civil society to work for the development of modern sustainable cities. In parallel, the fast-growing economy and the long-time investment in superior infrastructures all over the world are supporting sustainability during the normal operation time of the structures, and providing resilience under the effect of sudden disruptions and a quick recovery afterward. Sustainability, resilience, and recoverability (restorability) are three terms that were recently introduced to the field of civil engineering. The mutual relations between the three terms and structures need to be clearly established in order to identify the required sustainability, resiliency, and recoverability analysis methods, which in turn lead to the development of advanced design guidelines for sustainable, resilient, and recoverable structures.

This chapter provides a definition for the sustainability performance-based design of urban infrastructures, and briefly addresses the interrelation between the sustainability, resilience, and recoverability of structural systems. The global efforts made to enhance the resiliency of urban systems are presented in light of the design recommendations and considerations given by the available design codes. In addition, the academic literature concerning this endeavor is briefly presented in view of modern structural systems and the technologies proposed for the construction of resilient infrastructures. With respect to seismic forces as accidental actions, a mechanical model providing qualitative and quantitative measures for the seismic performance of urban structural systems is well-defined. The real situation of existing old and modern urban infrastructures is evaluated. Lastly, several models for the adoption of fiber-reinforced polymers (FRP) to enhance the recoverability/resiliency and control the performance of existing and modern reinforced concrete (RC) structures are comprehensively discussed.

6.2 SUSTAINABILITY DESIGN OF URBAN STRUCTURES

The holistic concept of sustainability depends on satisfying the needs of present and future generations, while creating a balance between society, environment, and economy direction. In addition, mutual influences between the three directions should be introduced so they can benefit from the positive aspects and keep negative impacts to a minimum. From the authors' point of view, although there are a number of experiences that can be judged as promising and encouraging toward the construction of sustainable cities, ensuring that the concept of sustainability is more widely applied remains challenging. The sustainability of infrastructure provides one of the crucial practical challenges for transforming ideas and principles into concrete realities. In the field of civil engineering, there are many studies adopting new, renewable, and smart materials to reduce the negative impacts of available/traditional construction materials on the environment and control construction costs, while raising the efficiency of the functionality of modern infrastructures. Moreover, numerous research deals with natural hazards and disasters associated with improper human actions to ensure the rapid recoverability/restoration of the functionality of critical infrastructures. Other studies have tremendously contributed to the development of effective health-monitoring methods and intelligent calculations for reliably following up on the states of structures. Bringing together the interactions among the scientific outputs available in abundance across many disciplines (structural, disaster prevention and mitigation, advanced materials application, information and communication technology, etc.) will have undeniable effects in bringing the design of a wide range of sustainable structures into reality.

6.3 MODEL FOR SUSTAINABILITY PERFORMANCE-BASED DESIGN OF RC STRUCTURES

The authors here propose a sustainability-based design model for modern RC structures. This model includes several analysis/design approaches and design-based requirements that constitute mutually influent interactions among several design objectives to satisfy the three major pillars of sustainability performance-based design, i.e., economical and durable, green and healthy, and safety. This model also articulates design properties (characteristics) and the corresponding assessment indices (metrics) of sustainable structures, as shown in Figure 6.1.

Green, light-weight, and life-cycle cost analysis/design approaches can result in the protection of natural resources, development of high performance but environmentally friendly materials, and reduction in construction and maintenance costs of modern RC structures. Resilience-based design mostly emphasizes developing all the necessary measures to provide structures with deliberated protection systems/elements/resources to meet the anticipated diversity of hazards/threats to ensure the long-term functionality of human-based constructions. The smartness of structures can be realized through the introduction of special components, special processing, and the adaptation of the materials-design. The longevity of a structure is dependent on the resistance of its components to harsh environmental conditions and natural hazards or human-made disasters. Therefore, the structural system should allow its components to be renovated, repaired, and replaced. That is, the longevity of a structure represents the outputs of the green-design, the light-weight design, and the smartness and resilience-based design requirements. The positive impact of any design area on sustainability performance-based design cannot, in general, be guaranteed without taking into account the triple bottom line of sustainability (economic, environmental, and society). For example, durable, resilient, and reliable structural systems (based on advanced technologies, highly durable materials, and strong resisting systems) can satisfy the sustainability-based performance in some ways, but economic assessments may be exaggerated. On the other hand, other less expensive solutions can be applied, but may significantly harm the built environment. The society's direction is directly or indirectly influenced by final decisions, and so mutual harmonization between findings/outputs at all design phases (planning and design, construction, and operation) is critical to balancing the three dimensions of sustainability.

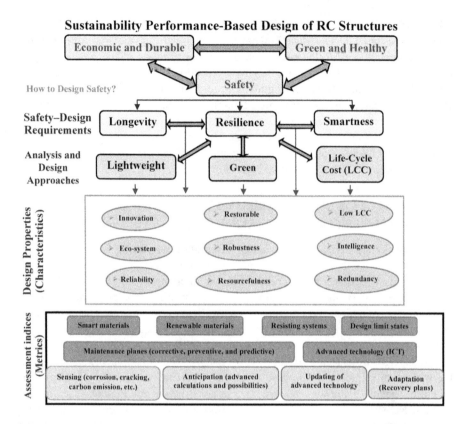

FIGURE 6.1 Proposed model for the sustainable performance-based design of structures.

To ensure the sustainable performance-based design of modern structures, design of sustainable structures should be characterized by resourcefulness, reliability, redundancy, recoverability/restorability, robustness, intelligence, innovation, being environmentally friendly (eco-system), and having a low life-cycle cost (LCC), as shown in Figure 6.2. **Resourcefulness** means that established systems are based on renewable resources of materials, information, technical elements, recovery and maintenance plans, and management strategies to achieve certain goals and maintain the integrity of the system so it can be continued indefinitely. **Reliability** is the assurance that all components of a structure must function effectively throughout both the usual operating cycles and the times of pre-predicted perturbation events. It also ensures that the structure has a safe exit when entering the stage of functional-loss in the event of an unexpected external shock. **Redundancy** refers to the adoption of a set of systems/resources that jointly or sequentially support continuity of structure functions under any interrupted circumstances.

The **Recoverability/restorability** of a structure is determined when it suffers controlled damage during an interruption, and it is provided with deliberated plans and a collection of resources after the exposure to accelerate the speed of recovery of the structure's functions. **Robustness** is a consistent principle in the design of the

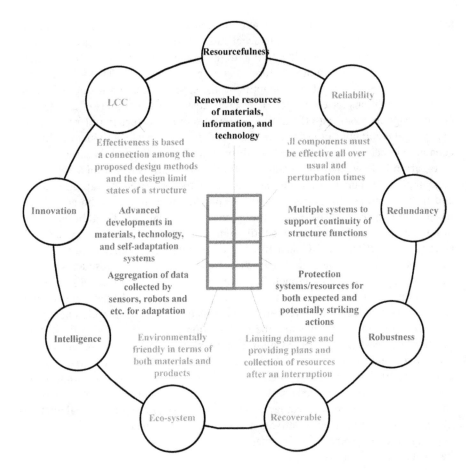

FIGURE 6.2 Design properties (characteristics) for the sustainable-based design of structures.

components of a structure according to its assigned function toward providing protection systems against what is expected to happen, taking into account everything that is new and may potentially change in the future. **Eco-system** is based on the construction of structural and infrastructure systems that are environmentally friendly in terms of both the materials adopted in the different components and products over the structure's lifetime, e.g., construction waste, production residues associated with structural function, and the impact of carbon emissions. The **intelligence** of a structure relies on the aggregation of data collected by all sensors that monitor the performance of all components for analysis allowing the highest possible efficiency during operation, saving energy, protecting all available resources, and ensuring constructive interaction between the various components. It predicts problems before they occur and provides advanced computational solutions to maintain functional efficiency over time. **Innovation** represents the extent to which advanced developments in materials, technology, and self-adaptation systems can be adopted by structures before and after construction. Following up on the continuous development in the field of energy-saving, the development of durable/cost-effective/light-weight building materials and innovative ways to accelerate construction and increase products' efficiency have been created. In addition to the tremendous progress in digital data science to track the health of structures, many innovations associated with the creation of new systems capable of adapting to the environment and its unpredictable changing conditions have emerged. **Life-cycle cost** links the outputs of all the proposed design methods with the design limit states of a structure in order to achieve the optimal continuous use of the structure and its components without any adverse economic impact of design limit states throughout the life-span of the structure.

Figure 6.1 shows that the sustainability of a structure is based on a very complex interaction between the characteristics of the proposed design approaches and the design-based requirements. This confirms that they must be completely interdependent in order to realize the required sustainability. In addition, in the design stage, the construction stage, the life-cycle stage, and the emergency stage (interruption actions) of RC structures, a large number of indices are necessary to assess the performance of structures designed using the proposed sustainable performance-based design model. Some of these indices are listed in Figure 6.1, such as the adoption of smart materials, the replacement ratio of natural resources with materials of renewable resources, the added value of renewable energy, the number of resistance systems and their interactions, acceptable design limit states under life-cycle loads, accepted maintenance planes (corrective, preventive, and predictive), and the application of advanced technology (ICT) in construction and during the life-cycle stage. In addition, in the life-cycle phase, other indices are very critical, such as the success of the applied health-monitoring system in tracking structural performance changes; the accurateness of the adopted calculations and possibilities to correctly activate mandatory maintenance, the readiness of the application of recovery plans in emergencies, and the applied technology should be evaluated for improvements.

6.4 SUSTAINABILITY, RESILIENCY, AND RECOVERABILITY

The proposed model shows that restoration/recoverability is a necessary property of structures to ensure rapid recovery of system functionality after non-traditional events and thus it serves as a benchmark for the resiliency of structures, which in turn is a direct indicator of the performance of a sustainable structure at the time of interruption.

Assuming a structure is affected by an immense interruption, Figure 6.3 shows the interrelation between recoverability, resilience, and sustainability. When the strike is an anticipated action, and structures return to full operational capacity quickly after the interruption, this would be a case of quick recovery and the system can be seen to be resilient, and the structure sustainable. In cases where extreme natural hazards affect structures, the recovery plans should be adequate enough to restore functionality within a short time and thus the system is considered partially resilient and the structure is sustainable. For structures that may be severely damaged, no recovery plan can be adopted to restore the structure's functionality, which refers to a non-resilient system and unsustainable structure.

Resiliency and Recoverability of Concrete Structures

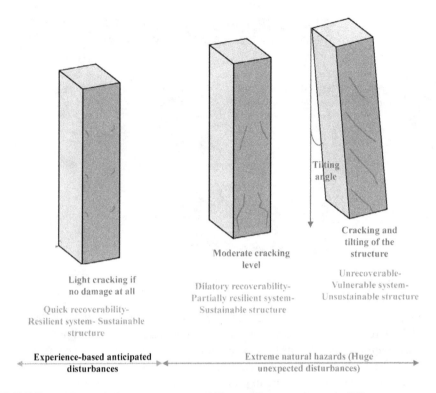

FIGURE 6.3 Interrelation among recoverability, resilience, and sustainability.

6.5 WHAT GLOBAL EFFORTS ARE BEING MADE TO ENHANCE THE RESILIENCY OF RC STRUCTURES TO MOVE TOWARD FULLY SUSTAINABLE COUNTRIES?

Severe earthquakes are one natural hazard that may strike structures at least once in their lifetime. In this case, the seismic resilience of infrastructure or structural systems is the capability of the structure system to withstand the effects of earthquakes and swiftly recover to its original functionality. The time required to restore/recover that functionality is a critical parameter of resilient structures. To generalize, the goal of resiliency to be applied more broadly, that is, to achieve national resiliency, quite a few countries have established different strategies that correspond to anticipated natural hazards and/or man-made disasters. For instance, in Japan, fundamental plans are being developed to achieve the nation's resiliency goals at the local and regional levels: by the national government for national resilience and by prefectures and municipalities for regional resilience. The fundamental plan includes a basic disaster prevention plan and national spatial strategies, as shown in Figure 6.4. The basic concept in this plan is the creation of a strong and resilient country. The basic policies include: protection of human lives; avoiding fatal damage and maintaining important functions of the nation and the society; minimizing damage to the property of the citizenry and public facilities; and swift recovery and reconstruction.

FIGURE 6.4 Japanese government plan for building national resilience.

In the United States, the National Earthquake Hazards Reduction Program (NEHRP) recommends seismic provisions, and presents the minimum recommended requirements that are necessary for the design and construction of new buildings and other structures for resisting earthquakes and ground motions. The objectives of these provisions are to provide reasonable assurance of seismic performance that will avoid serious injury and loss of life; preserve means of egress; avoid loss of function in critical facilities; and reduce repair costs, both structural and nonstructural, where practical.

6.5.1 THE INTERNATIONAL ORGANIZATION FOR STANDARDIZATION (ISO)

The International Organization for Standardization (ISO) developed different standards for the performance requirements of concrete structures. Environmental, economic, and social aspects are considered to ensure green and sustainable materials and structures; minimum life-cycle costs; and enhancement of safety and recoverability design. For example, ISO 15392 and ISO/TS 12720 were developed for sustainability in building construction, and they present general principles and their application, respectively. Other standards have been developed for construction works and construction products and services. In addition, ISO 19337 has been established for the performance and assessment requirements of design standards on structural concrete. One subsection of ISO 19337 defines restorability as the ability of a structure or structural element to be repaired physically and economically when damaged by the effects of considered actions. In the general requirements, a design standard for structural concrete shall be based on quantitative performance evaluation at the limit states. The design shall consider safety, serviceability, restorability, structural integrity, robustness, environmental adequacy, and durability.

The research community has also endeavored to define new resilient reinforced concrete structural systems, satisfying the requirements of both the limited resilience loss and the short time recovery of the original functionality. However, consolidating efforts in various disciplines of research and practical application is necessary to succeed in developing general plans for the design of resilient systems that allow application at the global level.

6.6 WHAT IS THE SITUATION OF CURRENT RC STRUCTURES?

In previous years, several populations inhabiting active earthquake zones around the world have been subjected to violent forces which have caused catastrophic consequences. Extensive research has been undertaken in this direction and has revealed the causes of the various levels of observed damages. Some of the observed damages are sketched in Figures 6.5(a–b) for the bridge piers and beam-column joints of RC buildings, respectively. Figure 6.5(c) shows the experimentally defined lateral load-displacement relationships for these structures. It is evident that the critical components of existing structures would suffer from shear failure, lap-splice failure, and/or flexural failure. All failures are related to under-designed reinforcement details or the deterioration of material quality over time.

The current design codes of structures exposed to seismic-loads adopt the philosophy of a limited damage level in the inelastic stage of the structures, which relates to a rapid recovery time. Fahmy (2010) studied four groups of rectangular RC bridge columns, available from literature and designed according to the available seismic design codes, to investigate their recoverability from experimentally simulated seismic forces. Results showed that the current provisions for the reinforcement details of RC structures could ensure a respectable resistance system with reasonable deformability; however, recoverability cannot be achieved after a drift of 2%. In addition, Fahmy (2010) showed that aged and under-designed structures need a large amount of work in order to introduce reasonable retrofitting systems to fulfill the requirements of the current codes. Generally, existing communities/cities all over the world should be updated for the required recoverability.

6.7 THE IMPORTANCE OF ENHANCING THE RECOVERABILITY OF RC STRUCTURES

Many existing structures represent historical, cultural, or economic value that cannot be disregarded. In addition, there are other structures with a high-level of occupancy throughout most of the day. These structures are vulnerable to severe seismic events, which may not have been taken into account at the time of implementation. Achieving the goal of sustainable structures in existing cities to realize all the characteristics of it being a modern sustainable city is prohibitively expensive. In order to contribute effectively to the required sustainability, it would be acceptable to upgrade structures based on different levels of controlled/limited damage in the inelastic stage consistent with the importance of the function and service provided by the structure and its importance in terms of economic and heritage value.

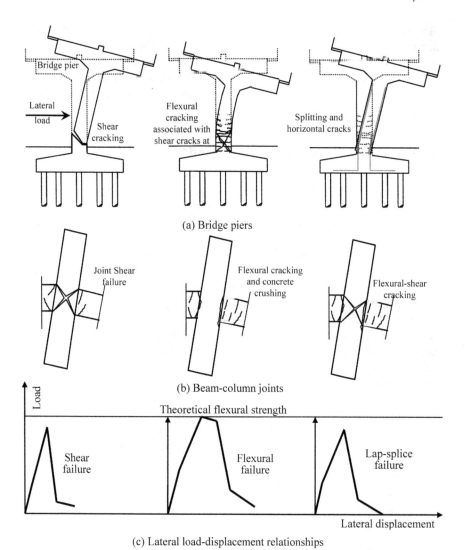

FIGURE 6.5 (a) Failure modes of old existing bridge piers. (b) Failure modes of beam-column joints of old existing buildings. (c) Load-displacement response of existing under-designed structures.

6.8 EFFORTS DONE TO ENHANCE RECOVERABILITY/ RESTORABILITY OF RC STRUCTURES

6.8.1 Modern Seismic Design Codes

Seismic design codes (JSCE 2000, Federal Emergency Management Agency (FEMA), AASSTO 2011, and CSA S6-06 2010) have been updated to include a definition of seismic performance design (PBD) for structures and infrastructures: a perfect default behavior is adopted by design codes and through good design taking into account exposure to earthquakes of varying density and probability of retune period, the importance level of the structure, and the geotechnical characteristics at the construction site, to create structures capable of achieving the functional objectives required. The Japanese code considers three limit states: structure safety limit, serviceability limit, and the restorability limit, as shown in Figure 6.6. At present, only the restorability (repairability) limit state under an earthquake action is defined in the Japanese standards. Other damage deteriorations, such as environmental actions, have no specific definition. In the future, it is recommended to consider the restorability limit state under different specific actions, such as fatigue, durability, and fire. In addition, no specific design guideline has been defined to satisfy the provisions of the required resiliency (Table 6.1).

The Federal Emergency Management Agency of the Applied Technology Council (ATC) and the Building Seismic Safety Council (BSSC) collaborated on the development of the NEHRP Guidelines and Commentary for Seismic Rehabilitation of

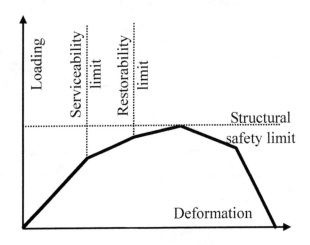

FIGURE 6.6 Limit states of laterally loaded structures. (From JSCE 2000.)

TABLE 6.1
Limit States Adopted by JSCE (2000) for Seismic Performance Design

Structural Safety	Damage and large deformation should not affect the structure's integral stability. And, no threat to the safety of the user and communities' life	
	Limitation under specific actions	Fatigue Limit State (fatigue damage caused by recycle action)
		Durability Limit State (damage caused by environmental action)
		Fire Resistance Limit State (damage caused by fire)
Serviceability Limit State	Structural response should not affect the structural serviceability	
	Limitation under specific actions	Fatigue Limit State (fatigue damage caused by recycle action)
		Durability Limit State (damage caused by environmental action)
		Fire Resistance Limit State (damage caused by fire)
Restorability Limit State	Damage or local destruction can be repaired in the allowable range of economy, technology, construction period, that can achieve the request to continue to use	

Buildings [FEMA 274]. The main concept of these guidelines is the application of certain design requirements that correspond to predefined limits of damage, i.e., target performance level, to design a structure for earthquakes with different densities. Figures 6.7(a-b) show the global lateral load-displacement relationships of ductile and nonductile structures, respectively, and the performance levels (immediate occupancy, life safety, and collapse prevention) are defined in the same figures. Table 6.2 presents a description of structure damage corresponding to each performance level and the expected business interruption as well.

NEHRP Recommended Provisions for Seismic Regulations for New Buildings and Other Structures (FEMA-302) adopts two criteria: reduce the risks associated

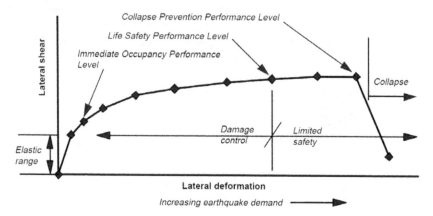

(a) Performance and structural deformation demand for ductile structures

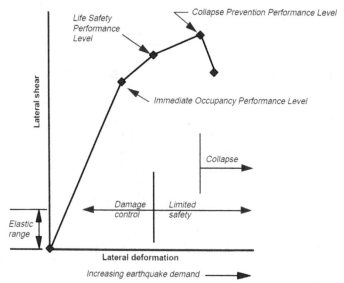

(b) Performance and structural deformation demand for nonductile structures

FIGURE 6.7 Limit states of laterally loaded structures. (From FEMA 274.)

TABLE 6.2
Building Performance Levels Per FEMA 273/274/356

Performance level	Damage description	Downtime
Immediate occupancy	Negligible structural damage; essential systems operational; minor overall damage	24 hours
Life safety	Probable structural damage; no collapse; minimal falling hazards; adequate emergency egress	Possible total loss
Collapse prevention	Severe structural damage; incipient collapse; probable falling hazards; possible restricted access	Probable total loss

with earthquakes to life for structures appropriate for their main function and use and improve the capacity of significantly important facilities and structures to continue operating during and after seismic actions. Adopting these provisions in the design of modern RC structures can result in repairable structures, but the impact on the economic side may be considerable. For earthquakes greater than design levels, these provisions adopt the criterion of a low probability of structural collapse.

6.9 APPLICATION OF FRP COMPOSITES TO ENHANCE THE RECOVERABILITY OF RC STRUCTURES

There is a call to upgrade the structural performance of a large number of existing structures to realize supreme seismic performance with or without a slight increase in cost compared with those required in ordinary structures. Hence, the key concept in this section is to introduce retrofitting techniques that can satisfy the characterizations of sustainable seismic performance design to enhance and exceed the advantages of conventional structures. If a substantial increase in cost is not required, then it is rational to design structures that suffer no damage or slight damage from very rare earthquakes, and such designs will be accepted by society. As a result of the enhancement of the seismic performances of individual structures in a city, the performance of the city itself will be improved remarkably.

Fiber-reinforced polymers are advanced composite materials that have several advantages and which has led the research community to examine its application in the strengthening/retrofitting of existing structures. Advanced composite materials can meet many of the required sustainability characteristics: additional resources for construction materials along with the available traditional materials, solving the longevity problem through superior resistance to corrosion and fatigue loadings, and ensuring a recoverable response due to the elastic behavior of the fibers to rupture, as shown in Table 6.3. Moreover, advanced composites include various types such as carbon FRP (CFRP), glass FRP (GFRP), aramid FRP (AFRP), and basalt FRP (BFRP), which cover a wide range of mechanical characteristics, e.g., high modulus, high ductility, etc. That means that the diversity of FRP types and their mechanical properties are able to reach the highest efficiency at the lowest possible cost.

TABLE 6.3
Mechanical and Physical Characteristics of Fiber Reinforced Polymers (FRP)

Types	Strength (MPa)	E (GPa)	Density (g/cm³)	Stress-strain	Corrosion resistance	Fatigue resistance
Steel	310	210	7.8		bad	low
CFRP	3500	230	1.78		good	excellent
BFRP	2100	80	2.65		good	good
GFRP	1500	70	2.55		normal	good

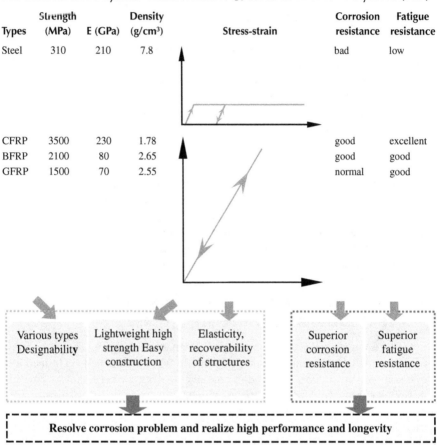

6.10 RECOVERABILITY OF EXISTING STRUCTURES USING FRP COMPOSITES

It is well known that by confining a concrete element in two out of three mutually perpendicular directions the ultimate compressive strength of the element in the third direction increases considerably. The effectiveness of confining FRP to concrete, generally, initiates after the yielding of the inner steel stirrups through a continuous increase in the resistance of the lateral expansion of the concrete due to its elastic behavior leading up to failure, see Figure 6.8(a). When an FRP jacket acts in the direction of transverse reinforcement, it contributes to the shear resistance components as well. Furthermore, a continuous increase in fiber restraint to the outward movement of the confined concrete has another positive impact on the relative slippage between short lap-splice reinforcement, which is accompanied by

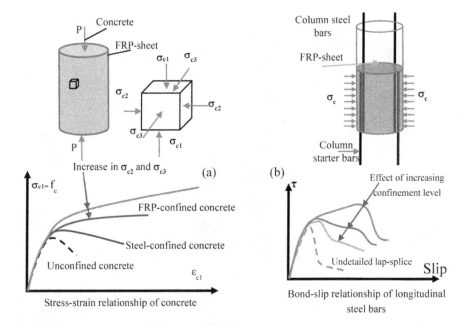

FIGURE 6.8 FRP confinement and (a) Stress-strain relationship of compressed concrete. (b) Bond-slip relationship of the longitudinal steel bars.

a significant overall improvement of the bond-slip relationship, as shown in Figure 6.8(b). In conclusion, an external FRP jacket is a passive seismic-resistant tool that can shift the brittle failure modes of existing structures to a more ductile behavior through a considerable increase in concrete compressive strength and corresponding strain capacity, a contribution to the shear resistance mechanism, and the control of the load-slip relationship of under-designed lap-splice reinforcements. Near-surface reinforcements have been successfully articulated by numerous researchers as being in direct correlation with the observed increase in the flexural resistance of many structural elements. Therefore this is a promising technique for the retrofitting of flexural-deficient structures. Hence, designers can apply both FRP jacketing and NSM techniques to work on the performance of the structure, i.e., deformability and flexural resistance, respectively.

Figure 6.9(a) schematically shows FRP retrofitting techniques using FRP jacketing or NSM FRP reinforcement as well as external FRP jacketing. Three mechanical models are presented in Figure 6.9(b) for comparison between the performance of an existing structure and the performances of two structures retrofitted with FRP jackets and NSM FRP reinforcement and FRP jackets. FRP composites in the transverse direction ensure an appreciable increase in the inelastic deformability of the retrofitted structure and maintain the flexural strength equal to or slightly higher than the theoretical strength, with a safe exit through a gradual degradation of the strength; however, robustness of the system may not be guaranteed as the residual strength would be lower than the yielding strength of the structure. On the other hand, the other system shows that the NSM FRP rebars can provide a design-controllable

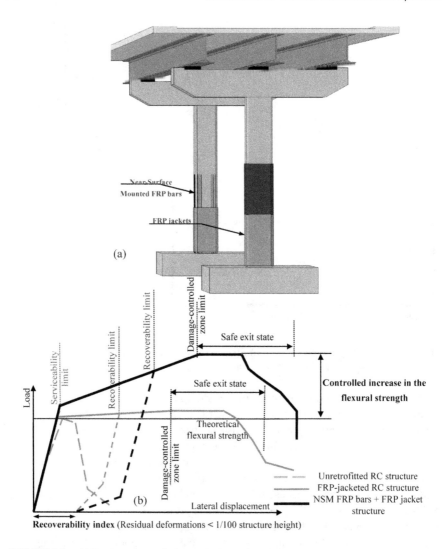

FIGURE 6.9 (a) Proposed retrofitting techniques for existing structures using FRP composites. (b) Lateral response of unretrofitted columns versus FRP retrofitted columns using two different techniques.

post-yielding stiffness after steel yielding, and the decrease in the lateral resistance at failure may experience a reasonable resistance.

The performance of the retrofitted structure is characterized by three limit states: serviceability, damage controllable, and ultimate states. The width of the serviceability state is dependent on steel yielding, and the limit of the damage controllable state corresponds to the maximum achieved flexural strength. The structure can be fully operational before steel yielding, and limited repair is necessary up to the recoverability limit that corresponds to a permanent deformation less than or equal to 1% of the structure height; heavy repair is compulsory after the recoverability limit, and reconstruction of some parts or demolishment of the whole structure could be the cost-effective choice when degradation of strength cannot be controlled, Figure 6.9(b). The use of additional longitudinal FRP rebars together with the external FRP jacket can be associated with an appropriate shift in the recoverability limit and the limit of the damage controllable state, Figure 6.9(b).

Figure 6.10 shows the test results of some specimens tested before retrofitting with FRP composites, the response of a column retrofitted with a FRP jacket, and another retrofitted with both NSM BFRP rebars and BFRP sheets.

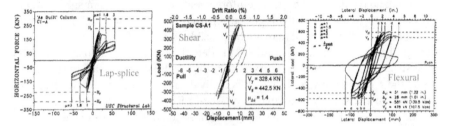

(a) Seismic performance of existing under-designed structures

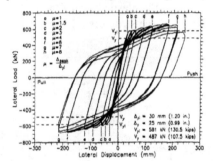

(b) Seismic performance of FRP-jacketed column, Seible et al. [1997]

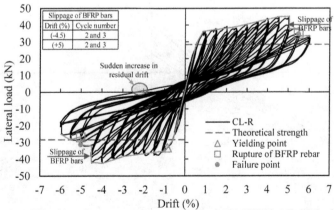

(c) Seismic performance a column retrofitted with NSM basalt FRP rebars and basalt FRP-jacketed, Fahmy and Wu [2016]

FIGURE 6.10 Lateral load-displacement relationships of unretrofitted and FRP retrofitted columns.

6.10.1 Limit States of Existing Structures after Retrofitting Using FRP Composites

A comprehensive evaluation of the recoverability of existing bridge columns retrofitted with an external *FRP confinement* was done by Fahmy (2010). The enhancement in the post-yielding stage was mainly dependent on the improvement of the local axial stress-strain compression behavior of the concrete and/or the enhancement of the bond strength between the lap-spliced steel reinforcement.

After the yielding of the steel reinforcement, the level of damage gradually increases to different levels that may be repaired within a short or long time, but with other levels of damage it may not be possible to repair. In some circumstances, although the level of damage is limited, the demolition of structure members or the whole structure would be mandatory, e.g., tilting of the structure due to soil failure beyond the recoverability limit, which is defined as not exceeding 1% of the height of the structure. Using the available database of FRP-confined RC columns, Fahmy et al. (2010) correlates the relationship between residual drift ratios and drift ratios, as shown in Figure 6.11. The authors concluded that residual inclination of the structure due to induced damage in the plastic hinge zone may prevent the possibility of recovering the functionality of the structure when the drift demand exceeds 3.5%.

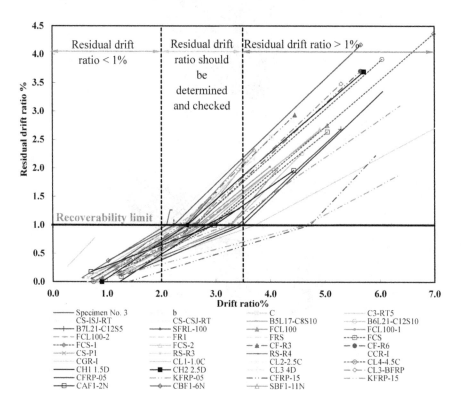

FIGURE 6.11 Column drift ratio versus residual drift ratio of a large database of FRP-confined RC columns.

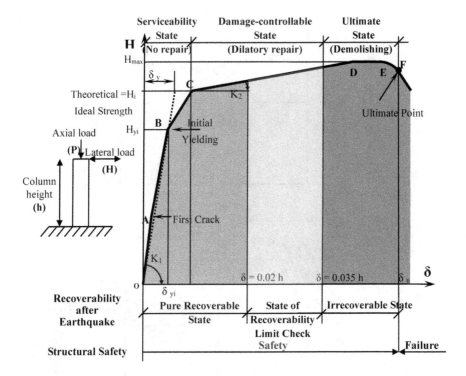

FIGURE 6.12 Recoverable and irrecoverable states of damage controlled FRP-RC columns. (Fahmy, 2010.)

For structures with a demand drift less than or equal to 2.0%, any tilt has no effect on the required recoverability. Within this range of 2.0–3.5% drift requirements, the residual deformations should be examined to make the correct decision regarding the recovery of the structure's functionality.

Figure 6.12 presents a complete overview of the lateral response of FRP retrofitted structures, which is classified into different damage-states in association with the findings related to the residual inclination (permanent deformation) as a measure for the possibility of recovering structure functions. The limit of the serviceability state corresponds to the global yielding of the steel reinforcement, the limit of the damage controllable state at 3.5% drift capacity, and beyond that the ultimate state. After a seismic hazard, the structure is fully operational before steel yielding; repair works are necessary up to a drift ratio demand of 2.0%. For the demand levels thereafter, residual inclination should be reviewed/determined before deciding on the required recovery strategy. When drift demand exceeds 3.5%, the structure is no longer functional or it may be close to collapse.

A numerical study was conducted on a RC column tested by Paultre et al. (2016) under the effect of seismic input, i.e., EL CENTRO earthquake motion, using Open System for Earthquake Engineering Simulation (Open SEES) software (Mazzoni et al.). The study included three columns: one serving as a reference for existing deficient structures, and the other two columns were retrofitted with a CFRP jacket or

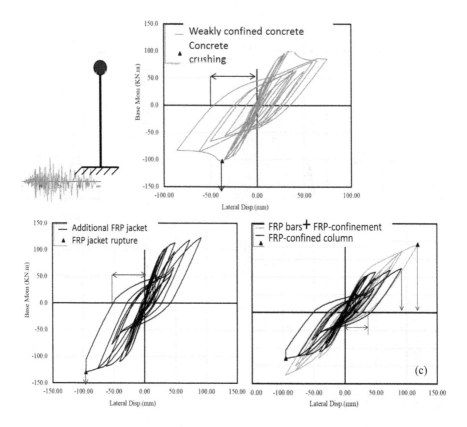

FIGURE 6.13 Dynamic results of (a) unretrofitted RC column, (b) retrofitted column with FRP-confined, and (c) retrofitted column with NSM FRP bars and FRP confinement.

NSM BFRP rebars and a CFRP jacket. The results are shown in Figure 6.13, which confirms an enhancement in the response using FRP confinement, and the superior performance is expected when additional FRP longitudinal reinforcement is considered to ensure a significant increase in the flexural resistance with a preferable mitigation of the residual deformations.

6.11 MODERN RECOVERABLE FRP-STEEL REINFORCED STRUCTURES

In conventional RC structures, it is no longer possible to control post-yield stiffness when steel is applied to the main reinforcement because the ratio between the steel hardening stiffness and its elastic stiffness is very small (1–2%). This ratio, in addition to well-confined concrete, can ensure a somewhat stable hysterical response after steel yielding but without a significant increase in the lateral capacity. In addition, the need for longevity in modern structures may not be possible using traditional steel reinforcements. Steel and FRP composites can be successfully combined together with different details that primarily serve the goal of sustainability and resiliency. Hence, the key concept in this section is to introduce rational structure systems for modern structures, which suffer controlled damage from simulated seismic forces; meanwhile a substantial increase in cost is not required to not only enhance but also exceed the advantages of conventional structures.

6.11.1 INNOVATIVE STEEL FIBER COMPOSITE BARS (SFCB)

Steel fiber composite bars (SFCBs) were proposed by Fahmy et al. (2010) for application as innovative reinforcements in modern structures located in seismic zones. The innovative reinforcement is a hybrid product consisting of a steel bar hybridized in a longitudinal direction with any of the advanced composite materials available (FRP); Figure 6.14(a). Mechanical properties of the new hybrid reinforcement outweigh the performance of the individual components because it combines the advantages of each material in a single structure. The external FRP protects the internal steel from harsh environmental effects, ensures the existence of controlled hardening after steel yielding, reduces the steel content in the entire structure, minimizes residual deformation, and reduces the construction costs of replacing steel bars with FRP rebars. On the other hand, the inner steel core ensures a higher elastic stiffness than that of the FRP rebar and allows the redistribution of induced stress when the external fibers rupture, as shown in Figure 6.14(b). That is, the proposed hybrid reinforcement supports longevity, resourcefulness, redundancy, cost-effectiveness, and resiliency for modern structures. The proposed innovative reinforcement can be adopted as longitudinal and transverse reinforcement [Figure 6.14(c)].

This innovative reinforcement was tested by Fahmy et al. (2010) as a longitudinal reinforcement for RC square bridge columns, and recently Ibrahim et al. (2016) examined the low cyclic fatigue response of circular columns reinforced with SFCBs and hybrid steel-FRP stirrups. Both studies confirmed the effectiveness of the proposed hybrid reinforcement in alleviating the residual deformations during the inelastic stage, shifting the recoverability limit, extending the plastic hinge length with a lower damage level, and ensuring a considerable increase in the lateral resistance compared to their counterparts of conventional RC columns, as shown in Figure 6.15.

6.11.2 FRP BARS AS LONGITUDINAL REINFORCEMENT AND STEEL REINFORCEMENT

The hybridization of FRP composites with steel bars is effective with small diameter bars for producing SFCBs; however, the increase in the steel bar size cannot guarantee the required hardening effect after yielding unless an excessive amount of fiber is used, which is not a practical solution. Nevertheless, diversity in FRP products is an incentive for proposing alternative solutions and introducing structural elements that meet all sustainable design requirements. Ibrahim et al. (2015) proposed new details for the reinforcement of bridge piers using steel reinforcements and FRP composites: the cross-section is longitudinally reinforced with steel and FRP bars and is transversely reinforced with both internal steel stirrups, which are relatively lower than those determined by the current design codes, and external FRP sheets which compensate for the shear and confinement demands, as shown in Figure 6.16(a). The study aimed to experimentally achieve a targeted seismic response, i.e., a ductile

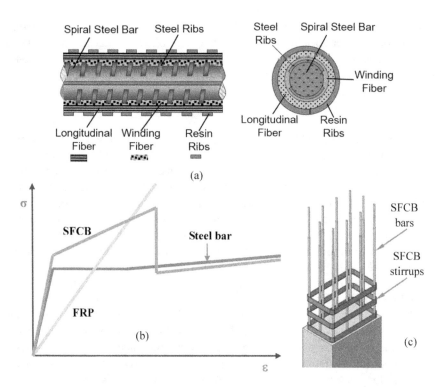

FIGURE 6.14 (a) SFCB geometry details. (b) Mechanical behavior of the SFCB versus steel and FRP bars. (c) Application of SFCB bars and stirrups as the longitudinal and transverse reinforcement of concrete columns.

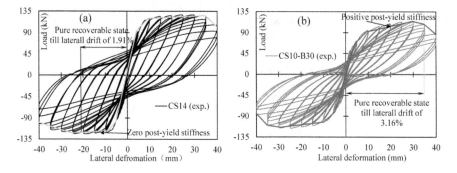

FIGURE 6.15 Lateral load versus lateral deformation of (a) a concrete column reinforced with steel bars and (b) a concrete column reinforced with SFCBs. (From Fahmy et al., 2010.)

damage controllable structure, through different bond conditions between FRP bars and the surrounding concrete, as shown in Figure 6.16(b). The test results of the two columns that successfully realized the required recoverability and controllability are shown in Figure 6.17.

Grids made of composite materials provide a low mass with a high rigidity and strength and are able to compete with traditional composite laminates. Furthermore,

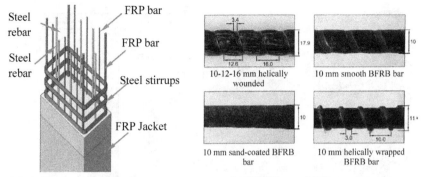

(a) Reinforcement details of concrete column using steel and FRP composites

(b) BFRP bars of different textures

FIGURE 6.16 A concrete column reinforced with internal steel and FRP reinforcement and external FRP jacket.

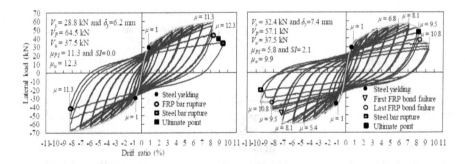

FIGURE 6.17 A concrete column reinforced with internal steel and FRP reinforcement and external FRP jacket.

since the orthogrid effectively resists normal forces in two orthogonal directions, the FRP grid can provide an increase in bending resistance when FRP is well anchored in adjacent components. So, it is also possible to use FRP grids in seismically active areas as an outer reinforcement for RC columns in addition to the internal steel reinforcement, as shown in Figure 6.17. This system can also ensure a ductile damage controllable performance for modern concrete structures reinforced with both steel and FRP (Figure 6.18).

6.11.3 Bond-Based Design Controllable FRP Bars as Longitudinal Reinforcement and Steel Reinforcement

FRP rods can be customized with special fabric details as a new design tool to control the level of damage and the global seismic response of RC structures. The fabric of these rods is designed to achieve a specific correlation relationship between FRP rods and concrete. Ibrahim et al. (2016) conducted an experimental and numerical study of the bond-based lateral response of reinforced concrete columns with steel

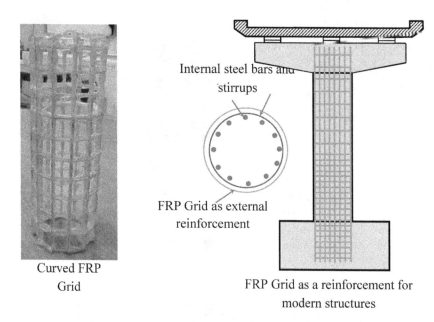

FIGURE 6.18 (a) FRP grid in a curved shape. (b) Concrete column reinforced with internal steel reinforcement and external FRP grid.

bars and FRP rods. Assuming different ranges of the characteristics of the bond-slip relationship of FRP rods, several important recommendations were made for the production of these new predesigned FRP rods. Significant results corresponding to different bond conditions on the lateral resistance of RC bridge columns are shown in Figure 6.19(a). Figure 6.19(b) shows the studied bond conditions between the FRP rods and the surrounding concrete, and Figure 6.19(c) shows the idealized bond-slip relationship.

Obviously, the lateral response of RC columns reinforced with steel and FRP bars is very sensitive to the FRP bond conditions. Precisely, the inelastic response (post-yield stiffness, stability response, and strength degradation) depends on the bond behavior of FRP bars. The ultimate bond strength is directly proportional to the slope of the post-yielding stiffness in the lateral load-displacement relationship: the stronger the bond, the higher the inclination slope after the yielding. The endpoint of this part depends on the strength of the bond, as shown in the results of Cases 1–3. The structure's ability to show a persistent resistance to the lateral deformations is mainly based on the width of the plateau zone of the bond-slip relationship, as shown from the results of Case 4. In order to ensure a very slow exit of the structure from its functionality, the increase in the fracture energy of the bond-slip relationship is very critical. Figure 6.19(d) shows the given recommendations for the bond-based design FRP bars for modern steel-FRP reinforced concrete structures (Ibrahim et al. 2016).

FRP composites have been used as longitudinal and transverse reinforcements in modern RC buildings located in seismic zones. Limited studies have been conducted, but it is claimed that FRP reinforced columns, beam-column joints, and moment resisting frames (MRFs) have an acceptable seismic response. However, the

Resiliency and Recoverability of Concrete Structures

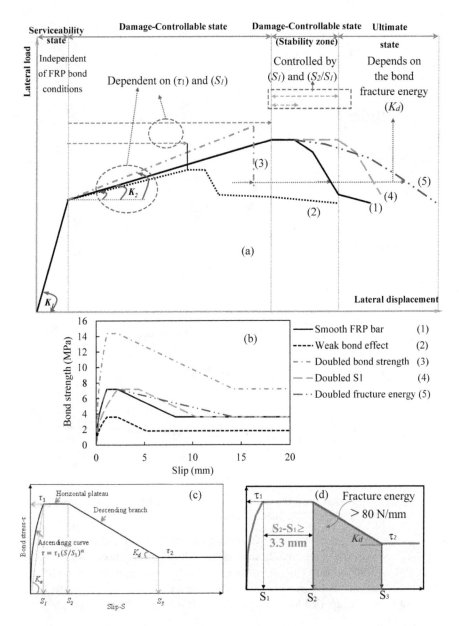

FIGURE 6.19 (a) Bond-based lateral response of RC columns reinforced with steel and FRP bars, (b) bond-slip relationship of FRP bars for the studied five cases, (c) idealized bond-slip relationship, and (d) recommended bond-slip relationship for bond-based designed FRP bars.

low stiffness of FRP composites is associated with a significant increase in demand for deformability (Figure 6.20).

Steel and FRP reinforcements have recently been introduced by Ibrahim et al. (2018) for the beam-column joints of modern RC MRFs, in which steel longitudinal bars are used in critical zones along with the longitudinal and transverse FRP

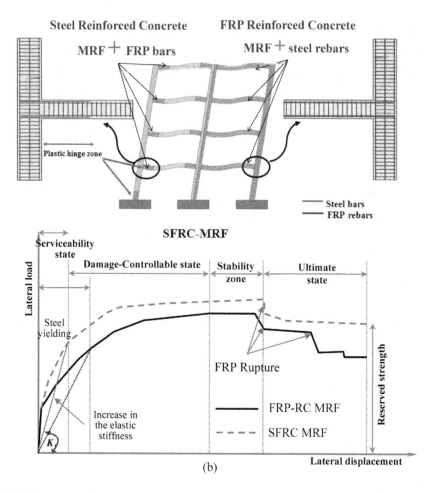

FIGURE 6.20 (a) Proposed reinforcement details for SFRC-MRF. (b) Response of FRP-RC MRF versus SFRC-MRF (3D FE numerical results).

composites. Partial replacement of FRP reinforcement in the critical zones with steel reinforcement provides a reasonable solution to reduce construction costs (reducing the amount of FRP reinforcement), to ensure robustness (lateral structure stiffness and resistance dramatically increase), and to realize redundancy (at failure steel reinforcement resists the applied load and brings a safe exit from functionality).

6.11.4 Innovative Resilient Systems Using FRP Composites

For more than two decades, assembling prefabricated concrete parts has been implemented using the prestressing technique; this system offers advantages by contributing to a significant increase in the quality of the product while achieving higher execution speed. These advantages have raised the interest of the research community for supporting its practical application in the construction of seismic-resistant systems. This system depends on the rocking behavior between column/

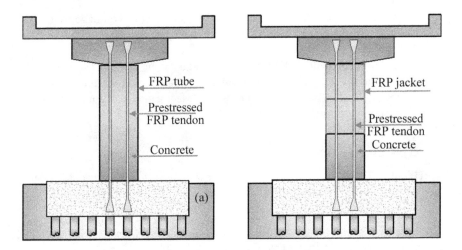

FIGURE 6.21 Application of FRP composites in the construction of precast segment bridge piers: (a) Concrete filled FRP tube and (b) FRP-confined concrete segments.

beam segments and adjacent structural elements, and the balance between the tension force in the prestressed tendons and the concrete compression resistance is responsible for the definition of the lateral strength. Failure is usually a tangible concrete crush at the most compressed column face. This is usually followed by a gradual decrease in the lateral strength after the peak load that may correspond to a low drift capacity. In order to enhance the performance of these types of structures, FRP composites can be adopted in several forms that ensure all characteristics of sustainable design. For instance, a FRP tube can be applied as external reinforcement in order to limit the concrete's outward deformability, thereby delaying early degradation of the structure strength; Figure 6.21(a). Multiple concrete segments externally confined with FRP sheets; see Figure 6.21(b), can significantly reduce construction costs by carefully designing FRP amounts for each segment. Furthermore, the traditional prestressing technique using steel strands can be replaced with FRP tendons, which can provide the required longevity and resiliency on the occurrence of anticipated hazards; see Figure 6.21.

6.12 CONCLUSIONS

1- A new sustainability performance-based model is proposed for RC structures, which are located in seismic zones. Design approaches and the corresponding characteristics are defined, and several indices measuring/evaluating the impacts of the proposed design model on the three pillars of sustainability are briefly discussed.
2- Experimental and numerical studies on the application of FRP composites in existing and modern structures in the light of the characteristics of suitability performance-based design are comprehensively presented.
   - Structural recoverability should be a key index for structural resilience as it is also an index of structural sustainability.

- A combination of FRP and steel can be used to realize damage controllable and restorable seismic structures with durable, smart, and green behaviors.
- Damage controllable FRP bars should and can be designed.
- Innovative systems combining modern construction techniques with the application of FRP reinforcement present a new resilient system for modern sustainable structures.

REFERENCES

AASHTO, *AASHTO Guide Specifications for LRFD Seismic Bridge Design*, 2nd Edition. American Association of State Highway and Transportation Officials, Washington, DC, 2011.

BSSC Program on Improved Seismic Safety Provisions, NEHRP (National Earthquake Hazards Reduction Program) Recommended Seismic Provisions for New Buildings and Other Structures: Provisions, Vol. 302, 2003 FEMA.

CSA, *Canadian Highway Bridge Design Code (CAN/CSA S6-06)*. Canadian Standards Association (CSA), Toronto, ON, 2010.

Fahmy, M.F.M., Enhancing recoverability and controllability of reinforced concrete bridge frame columns using FRP composites, Civil Engineering, Ibaraki University, Japan, 2010.

Fahmy, M.F.M. and Wu, Z.S., Retrofitting of existing RC square bridge columns using basalt FRP rebars, CD-ROM, 6th International Conference on FRP Composites in Civil Engineering (CICE2012), Rome, Italy, June 13–15, 2012.

Fahmy, M.F.M. and Wu, Z.S., Exploratory study of seismic response of deficient lap-splice columns retrofitted with near surface–mounted basalt FRP bars, *Journal of Structural Engineering*, 04016020, 2016. doi:10.1061/(ASCE)ST.1943-541X.0001462.

Fahmy, M.F.M., Wu, Z.S., and Wu, G., Seismic performance assessment of damage-controlled FRP-retrofitted RC bridge columns using residual deformations, *Journal of Composites for Construction, ASCE*, Vol. 13, No. 6, pp. 498–513, 2009.

Fahmy, M.F.M., Wu, Z.S., and Wu, G., Post-earthquake recoverability of existing RC bridge piers retrofitted with FRP composites, *Construction and Building Materials*, Vol. 24, No. 6, pp. 980–998, 2010a.

Fahmy, M.F.M., Wu, Z.S., Wu, G., and Sun, Z.Y., Post-yield stiffnesses and residual deformations of RC bridge columns reinforced with ordinary rebars and steel fiber composite bars, *Journal of Engineering Structures*, Vol. 32, pp. 2969–2983, 2010b.

Federal Emergency Management Agency (FEMA), NEHRP commentary on the guidelines for the seismic rehabilitation of buildings, Report No. FEMA 274, prepared by the Applied Technology Council for FEMA, Washington, DC, 1997.

Ibrahim, A.I., Wu, G., and Sun, Z.Y., Experimental study of cyclic behavior of concrete bridge columns reinforced by steel basalt-fiber composite bars and hybrid stirrups, *Journal of Composites for Construction*, Vol. 21(2), 04016091, 2016.

Ibrahim, A.M., Fahmy, M.F.M., and Wu, Z.S., 3D finite element modeling of bond-controlled behavior of steel and basalt FRP-reinforced concrete square bridge columns under lateral loading, *Composite Structures*, Vol. 143, pp. 33–52, 2016.

Ibrahim, A.M.A., Wu, Z.S., Fahmy, M.F.M., and Kamal, D., Experimental study on cyclic response of concrete bridge columns reinforced by steel and basalt FRP reinforcements, *Journal of Composites for Construction, ASCE*, 2015. doi:10.1061/(ASCE)CC.1943-5614.0000614.

Ibrahim, H.A., Fahmy, M.F.M., and Wu, Z.S., Numerical study of steel-to-FRP reinforcement ratio as a design-tool controlling the lateral response of SFRC beam-column joints, *Engineering Structures*, Vol. 172, pp. 253–274, 2018. doi:10.1016/j.engstruct.2018.05.102.

International Standard ISO/TS 12720.

International Standard ISO 15392.

International Standard ISO 19338.

Mazzoni, S., McKenne, F., Scott, M.H., Fenves, G.L., et al. Open system for earthquake engineering simulation user manual version 2.1.0, Pacific Earthquake Engineering Center, University of California, Berkeley, CA. http://opensees.berkeley.edu/OpenSees/manuals/usermanual/.

Wu, Z.S., Fahmy, M.F.M., and Wu, G., Safety enhancement of urban structures with structural recoverability and controllability, *Journal of Earthquake and Tsunami*, Vol. 3, No. 3, pp. 143–174, 2009.

7 Urban Infrastructures Resilience Assessing
An Overview & New Resilience Evaluation Theory

Wael A. Altabey, Mohammad Noori, and Ying Zhao

CONTENTS

7.1	Introduction	109
	7.1.1 Resilience	110
	7.1.2 Big Data (BD)	110
	7.1.3 Data Mining (DM)	112
7.2	Challenges	112
7.3	Methodology	113
	7.3.1 Network Model	113
	7.3.2 Quantification of Resilience Capabilities	114
	7.3.3 The Resilience Measures	115
	7.3.4 Input-Output Model	116
	7.3.5 Input-Output Inoperability	116
7.4	Discussions	118
7.5	Conclusion	120
References		120

7.1 INTRODUCTION

The infrastructure systems envisioned for smart cities can be developed through three stages, as shown in Figure 7.1. Infrastructure 3.0 brings all parts of the infrastructure puzzle together and incorporates them into a single interdependent and reliable whole. Infrastructure 3.0 provides real-time optimization and incident handling across all domains. It allows us to adapt to the pressures of rapid urbanization, climate change, and other trends by utilizing advances in sensors, controls, and software to predict outcomes, take actions, and manage systems more effectively. For example, in an Infrastructure 3.0 world, smart buildings and the smart grid cooperate seamlessly to optimize energy consumption. Smart buildings take on surplus energy when it is cheap and plentiful, storing it for later and feeding it back into the grid when demand is high. Traffic systems become more user-friendly, integrating

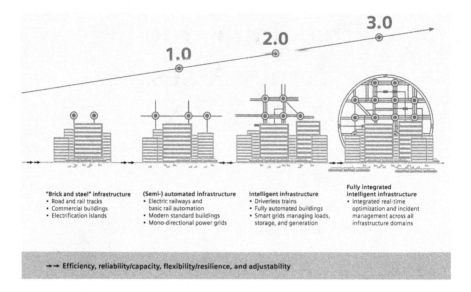

FIGURE 7.1 Stages of infrastructure development.

all transport modes and operators so that travelers can optimally plan their journeys using real-time information – which reduces both congestion and emissions. And command and control centers are capable of integrating transport, water, gas, and electricity networks to exercise pre-emptive actions or respond swiftly in a crisis (SIEMENS, 2018).

7.1.1 Resilience

Resilience has been used quite differently across a wide variety of social, technical, and economic disciplines. For example, the indicators used to assess psychological resilience differ so greatly from those used to assess infrastructure resilience, since psychological and critical infrastructure resilience are vastly different, while both are termed resilience, a focus on the indicators used to assess them both may show that they are two fundamentally different characteristics. In this regard, urban infrastructure resilience is the capacity of an urban infrastructure within a city to survive, adapt, and thrive, no matter what kinds of chronic stresses and acute shocks it experiences. The acute shocks include, for instance, earthquakes, wildfires, flooding, sandstorms, extreme cold, hazardous materials accident, severe storms and extreme rainfall, terrorism, infrastructure or building failure, and heatwaves. The flowchart of the framework set out for resiliency evaluation of infrastructures in smart cities is shown in Figure 7.2.

7.1.2 Big Data (BD)

Figure 7.3 shows examples of BD applications in smart cities. Smart city functions/operations generate very large amounts of data, while BD systems utilize this data

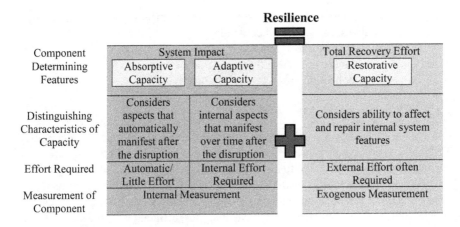

FIGURE 7.2 Resiliency evaluation of infrastructures.

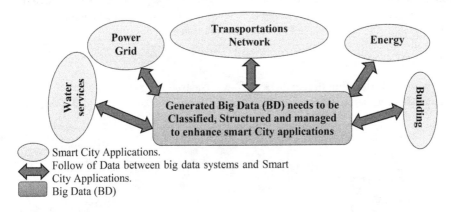

FIGURE 7.3 Smart city and big data relationship.

to provide information to enhance smart cities' harmonious operations. The BD systems will store, process, and mine data and information corresponding to various functions/systems in smart cities in an efficient manner to produce information to enhance different smart city services. In addition, the BD will foster decision making and accountability to plan for any expansion in either smart city services, resources, or areas. Furthermore, BD is extensive in terms of characteristics and features, like volume, variety, velocity, and/or variability. Overall, these features have three main characteristics (1, 2, and 3) and two additional characteristics (Fan and Bifet, 2013; Eiman et al., 2015):

- Volume: refers to the size of the data that has been created from all sources.
- Velocity: refers to the speed at which data is generated, stored, analyzed, and processed. Currently, emphasis is being put on supporting real-time BD analysis.
- Variety: refers to the different types of data being generated. Currently, most data is unstructured and cannot be easily categorized or tabulated.

- Variability: refers to how the structure and the meaning of the data constantly changes, especially when dealing with data generated from natural language analysis, for example.
- Value: refers to the possible advantages that BD can offer a business based on good BD collection, management, and analysis.

7.1.3 Data Mining (DM)

DM can be viewed as a result of the natural evolution of information technology. Thus, DM is a computer-based process or new computer-based technique for converting large sets of data (BD) to information and knowledge by finding patterns and opportunities within the data using different techniques of visualization, reduction of dimensionality, classification, and construction of models. Methods used in DM, as shown in Figure 7.4, come from statistical analytics, artificial intelligence (AI), such as Artificial Neural Network (ANN), Machine Learning (ML), Deep Learning (DL), etc., management science and information systems disciplines for pattern recognition, mathematical modeling, databases activities, and data management, including data storage and retrieval, and database transaction processing.

7.2 CHALLENGES

A method of assessing the resilience of an infrastructure and prioritizing improvements is considered important and timely. So an emphasis on building resilient

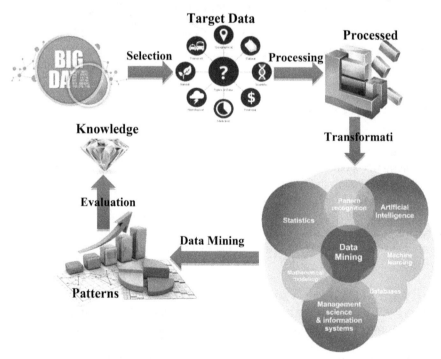

FIGURE 7.4 Data mining system.

Urban Infrastructures Resilience Assessing   113

structural systems requires that factors affecting resilience are be identified early in the development of a project. Opportunities for resilience can be lost if not planned and designed prior to implementation. On the other hand, infrastructure systems are becoming more complex and interdependent, but "smarter" in normal operation and use. However, there some basic challenges and questions that need to be clearly identified and addressed in this regard, such as:

1. Will these choices lead to "smarter" resilience when the infrastructure is exposed to extreme threats, such as extreme weather disasters or earthquakes?
2. As smarter systems are getting more complex, will they be more vulnerable?
3. How will these new "smarter" resilient systems prepare for, respond, adapt, and recover from extreme threats?
4. How can we properly evaluate resiliency, or how can we obtain the proper and practical metrics that provide indications of whether the resilient system is suitable for specific applications/conditions?

7.3 METHODOLOGY

Resilience in the context of civil and industrial engineering systems is usually expressed mechanistically as the ability of the system to bounce back after a major disturbance. This methodology has been most commonly applied to individual structures; however, we need a methodology that can be applied to assess structures connected in a networked lifeline. In the methodology introduced and discussed herein, we apply the concept and definition of engineering resilience to performance data obtained from networked systems. We then combine these measures with input–output models to relate resilience to interdependency.

7.3.1 Network Model

Our network model will be derived from systems of interdependent structures, as postulated by Nanjing city infrastructures. The systems are denoted by index values as follows:

1. Transportation networks, including bridges, tunnels, subways, railways, air travel, roadways
2. Transportation subsystems, including fueling/gas stations, mass transit, rail stations, and port facilities
3. Electric power delivery, with subsystems of distribution, transmission, and generation
4. Telecommunications, with subsystems of cable, cellular, internet, landlines, and media
5. Utilities, with subsystems of water supply, sewage treatment, sanitation, oil delivery, and natural gas delivery pipelines
6. Building support, with subsystems of HVAC, elevators, security, and plumbing

7.3.2 Quantification of Resilience Capabilities

Resilience is a dynamic multi-faceted process and its assessment should cover all evolutionary phases and essential system features (absorptive, adaptive, and restorative capability). Figure 7.5 shows a general illustration of system resilience. The x-axis signifies time and y-axis represents the measurement of performance (MOP) of a system. The first phase is the original steady phase ($t_0 < t < t_d$), in which the system performance assumes its target value. The second phase is the disruptive phase ($t_d < t < t_r$), in which system performance begins to drop (in most cases) due to disruptive event(s) at time t_d until it reaches the lowest level at time t_r. The third phase is the recovery phase ($t_r < t < t_{ns}$), in which the system performance starts increasing until the new steady level has been reached. During the second phase, the system's absorptive capability can be assessed by Robustness (R) combined with two complementary measures, Rapidity ($RAPI_{DP}$) and Performance Loss (PL_{DP}), to identify the maximum impact caused by disruptive events. The system performance loss can be interpreted and quantified as the region bounded by the graph of the MOP before and after the occurrence of negative effects caused by disruptive events, which can also be referred to as the system impact area.

During the third phase, the system's absorptive capability can be assessed by Robustness (R) combined with two complementary measures, Rapidity ($RAPI_{DP}$) and Performance Loss (PL_{DP}), to identify the maximum impact caused by disruptive events. The system performance loss can be interpreted and quantified as the region bounded by the graph of the MOP before and after the occurrence of negative effects caused by disruptive events, which can also be referred to as the system impact area. Time Averaged Performance Loss (TAPL) is introduced to encompass the time between the appearance of negative effects due to disruptive events up to full system recovery, and provides a time-dependent indication of both adaptive and restorative capabilities as responses to the disruptive events.

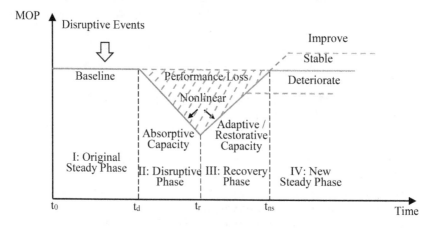

FIGURE 7.5 System resilience transitions and phases.

Rapidity ($RAPI_{RP}$) and Performance Loss (PL_{RP}) are developed to assess a system's adaptive and restorative capability. The newly attained steady level may be higher (resilience) or lower than (degradation, collapse) the previous steady state. A resilient system possesses the ability to recover its normal performance state level from a disruptive state, while a non-resilient system may gradually decline toward a low-performance level with a certain magnitude due to an unexpected event. Recovery Ability (RA) is quantitatively developed. Loss function, recovery function and fragility function were established in comparison with a mechanical analogy (Cimellaro et al., 2010).

7.3.3 The Resilience Measures

These are the key measures in the assessment of resilience.

Fragility: Lifeline resilience may be explored through the use of fragilities. Fragilities are tools commonly employed by structural engineers to characterize the probability of damage given a level of hazard demand such as wind velocity or ground acceleration. Most commonly derived for individual structures, in this investigation, we define fragility as dependent on a networked lifeline as a whole, i.e., we will connect the fragility equations between each individual index value in a network model to get the overall fragility equations of the network model.

Quality: Quality is a function derived by the Multidisciplinary Center for Extreme Event Research (MCEER) group and employed by many in the earthquake engineering community to describe structural performance over time following earthquakes. We extend this concept for each of the damage types in this work. In addition to earthquakes, we will address flooding, extreme cold, hazardous materials accident, severe storms and extreme rainfall, terrorism, infrastructure or building failure, to cite a few examples. For example, we apply the MCEER function to wind-induced damage in this chapter. In equation form, the quality $Q(t)$ is (ORourke, 2007; Dorothy, 2009):

$$Q(t) = Q_\infty - (Q_\infty - Q_0)e^{-bt} \quad (7.1)$$

Where Q_∞ is the capacity of the fully functioning structural system; Q_0 is post-event capacity; b is the parameter derived empirically from the restoration data following the event; t is time, in days, of post-events.

In addition, the integration of the area under the curve has been labeled Resilience (R) (ORourke, 2007). In equation form:

$$R = \frac{\int_{t_1}^{t_2} Q(t)dt}{(t_2 - t_1)} \quad (7.2)$$

Where t_1 and t_2 are the endpoints of the time interval under consideration. For the system infrastructure described in Section 7.1.1, we may evaluate resilience measures R_1 for subsystem (1) from Q_1, R_2 for system (2), etc., from post-event data.

We propose that the system's resilience in general for a set of total subsystems is a function of the individual system as follows:

$$R_S = g(R_1, \cdots, R_i, \cdots, R_n)$$

Where g() is a function to be determined that combines the individual resilience values in a way that reflects their interdependence and connectivity. It should be pointed out that the system resilience does not appear directly in quality equations, but indirectly incorporates the Rapidity and Robustness parameters of the individual subsystems.

7.3.4 Input-Output Model

Data collection and capture from sensors, users, electronic data readers, and many others from the systems of infrastructure described in Section 7.1.1, pose a most important challenge as the volume rapidly grows to become BD at inputs to resilience systems, for storing, organizing, and processing this data to generate useful results at the outputs of the resilience system. To further complicate the challenge, handling the interconnected communication infrastructures described in Section 7.1.1 to access contextual information in smart city applications and physical spaces to support good decision-making processes requires attention to various aspects of connectivity, security, and privacy (SDR, 2005). As the data comes from different sources with different formats, there is a need for advanced data management features that will lead to recognizing the different formats and sources of data, structuring, managing, classifying, and controlling all these types and structures. For example, most available data mining algorithms, as shown in Figure 7.4, are very suitable for big data mining applications like ANN, ML, DL, etc.

7.3.5 Input-Output Inoperability

We can combine BD with input–output models to relate resilience to interdependency by using Haimes (2004) derivation (3). Haimes was able to drive an interdependence model between various interconnected subsystems of the infrastructure in a smart city based upon the input–output economic model as follows:

$$x = Ax + C \qquad (7.3)$$

Where x is the vector of subsystem inoperability; A is the interdependency matrix between the various subsystems; and C is the disturbance or perturbation vector. The units for x are an inoperability or reduction of functionality.

The strategy of this derivation depends on the values of matrix A, where it takes on values between 0 and 1 and one interpretation is that a_{ij} is related to the probability of inoperability that the jth infrastructure contributes to the ith infrastructure. A value of a_{ij} of unity means that a complete failure of the jth infrastructure will lead to a complete failure of the ith infrastructure. A value of zero means that failure of the jth infrastructure has no effect on the ith infrastructure. For the 6-systems approach

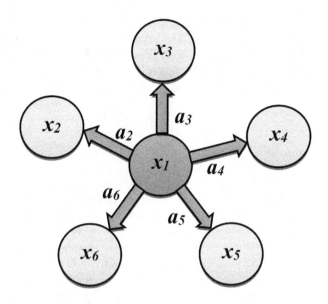

FIGURE 7.6 Selected matrix coefficients for the 6-systems model.

to infrastructure described in Section 7.1.1 we use the causal network diagram in Figure 7.6 to illustrate the relationship between a x_1 infrastructure (such as electric power delivery) central node and the other systems, such as transportation networks, telecommunications, transportation, etc. A value of a_{13} of unity means that a complete failure of the x_1 infrastructure system will lead to a complete failure of the x_3 infrastructure system.

The values of the C vector may be interpreted as the reduction in functionality or level of inoperability induced by extreme events such as hurricanes or earthquakes. Its values are bounded by zero and unity. For example, $c_1 = 0.8$ represents an 80% reduction in subsystem (1) operability or an inoperability of 80% as denoted by x_1 due to disruption. We must also note that derivation (3) components are steady-state positions of interoperability infrastructure when risk happens. A dynamic input–output model means restoration analysis within a time scale, i.e., the relationship between the recovery of the entire infrastructure and the time of recovery as follows (Haimes, 2004):

$$x = Ax + C + B\dot{x} \qquad (7.4)$$

Where the B matrix represents the relationship between x and its derivative with respect to time $\dot{x}$. Therefore, $\dot{x}$ represents the vector of the subsystem infrastructure's inoperability (RS) after part of the period of recovery time. The dynamic model gives rise to inoperability versus time curves of the form e^{-bt} during the recovery phase, from which resiliency R values may be estimated. Observed values of parameter b can be used to validate the dynamic model. Because of the differences in the restoration durations for each infrastructure, we compare the restoration using a normalized time scale.

In this work we have added two additional, and important, derivative terms for a dynamic input–output model for the development of resilience efficiency of urban infrastructures and the development of the description of the subsystems inoperability. These derivatives are the restoration response (RR), and restoration response indicator (RRI) as follows:

$$x = Ax + C + B\dot{x} + D\ddot{x} \tag{7.5}$$

$$x = Ax + C + B\dot{x} + D\ddot{x} + E\dddot{x} \tag{7.6}$$

The RR $\ddot{x}$ vector means the rate of change of the subsystem infrastructure RS of a restoration with respect to recovery time, where the D matrix represents the relationship between $\dot{x}$ and its derivative with respect to time $\ddot{x}$. This derivative is very useful for giving rise to RS versus time curves; for example, if the D matrix is a positive matrix in which all the elements are greater than zero, it means that the subsystem infrastructures RS is increased with the recovery time. If the D matrix is a negative matrix in which all the elements are less than zero, it means that the subsystem infrastructures RS is decreased with the recovery time. The RRI $\dddot{x}$ is very important to evaluate the resilience method used for the efficiency of urban infrastructures. RRI means the rate of change of the RR $\ddot{x}$ with respect to the recovery time, i.e., the RRI represents the exact description of the applicability and effectiveness of the resilience method used with urban infrastructures, where the E matrix represents the relationship between $\ddot{x}$ and its derivative with respect to time $\dddot{x}$. There are three profiles for RRI matrix E defined by the following:

If the E matrix is a positive matrix, it means the linear increase of RR at the positive direction of RR to the limit of final restoration.
If the E matrix is a negative matrix, it means the linear increase of RR at the negative direction of RR to the limit of final restoration.
If the E matrix is a zero matrix, it means the RR is regular or zero.

The first profile of RRI is considered the best one, where the resilience method used in this profile is positive, i.e., the capacity of urban infrastructures within a city to survive and adapt under the acute shocks experienced is positive.

7.4 DISCUSSIONS

As a means to illustrate the use of our methodology, we use the curve of restoration analysis with a normalized time scale (RS) for the power delivery system in the model for post-Hurricane Katrina landfall in Florida and the Hanukkah storm, 2006.(Dorothy et al., 2009; Graumann et al., 2005; FRHK, 2006; EOC, 2005). We used this data for the analysis of RR and RRI curves for a resilience method in each case. Figure 7.7 represents a comparison of the restoration in three different cities affected by the hurricane. As shown in Figure 7.7 the timescale is the time in days taken for restoration divided by the total recovery duration. Figure 7.8 represents a comparison between resilience ability applied in each city by considering the

Urban Infrastructures Resilience Assessing

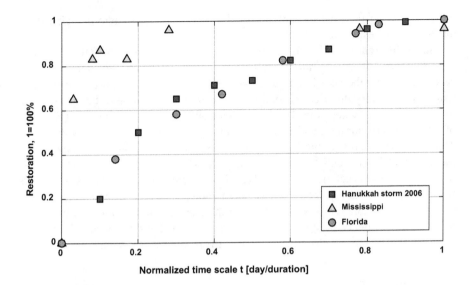

FIGURE 7.7 Restoration analysis with a normalized time scale.

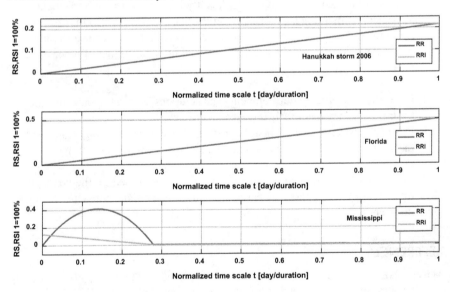

FIGURE 7.8 RR and RRI analysis with a normalized time scale.

two new derivatives for a dynamic input–output model that was presented in the proposed methodology. As shown in Figure 7.8, the RR of the resilience method in Hanukkah and Florida is positive. This means that the RS is increased with the recovery time, although the slope of RR in Florida is higher than the slope of RR for Hanukkah. This indicates that the increasing rate of RS with the recovery time in the Florida resilience system is more than the Hanukkah resilience system. Moreover, the RRI curves also confirm this point, i.e., the RRI of both cities is positive and constant, but the RRI for the Florida resilience system is higher than the Hanukkah resilience system.

This is different from the Mississippi resilience system, where the RR curve indicates that RS is increased during the first 17% of the total recovery duration, then decreased during the next 11% of total recovery duration and is constant during the remaining recovery duration. Also, this is clear in the RRI curve of the Mississippi resilience system, where the RRI is continuously decreased during the first 28% of total recovery duration and is constant during the remaining recovery duration. This is not acceptable in intelligent and resilient systems in smart cities, where it is required that RS is continuously increasing during recovery duration. This is the most important characteristic of a good resilient system in smart cities. As concluded from these observations, the RS curve is not enough to evaluate the resilience of a system in smart cities. It does not provide the adequate information necessary to describe the resilience or dynamic input–output model. Thus, we need RR and RRI derivation for a comprehensive evaluation of the resilience of systems.

7.5 CONCLUSION

The resilience of infrastructures will significantly improve with the implementation of a comprehensive methodology for risk and resilience assessment. This assessment is expected to lead to proactive innovations that, eventually, will raise the level of resilience of the infrastructure in smart cities. In addition, a number of other impacts are expected:

1. Fostering new product developments and solutions, generating new insights for infrastructures and their interdependencies
2. Providing novel tools and insights for rapid response planning, improved business continuity, and organizational adjustments for becoming more resilient
3. Enhancing the resilience of the society as a whole, based on the concepts of increased awareness, preparedness, and appropriate behavior during disasters

REFERENCES

Al Nuaimi, E., Al Neyadi, H., Mohamed, N., and Al-Jaroodi, J. (2015). "Applications of big data to smart cities," *Journal of Internet Services and Applications*, 41(5), 6–25.

Cimellaro, G. P., Reinhorn, A. M., and Bruneau, M. (2010). "Framework for analytical quantification of disaster resilience," *Engineering Structures*, 32(11), 3639–3649.

Emergency Operations Center (EOC) (2005). "Daily Service Outage Reports," August 29, 2005 through August 31, 2005, (Excel Spreadsheets for Each Day Summarizing Electric, Natural Gas, and Phone Outages) Louisiana Public Service Commission.

Fan, W., and Bifet, A. (2013). "Mining big data: Current status, and forecast to the future," *ACM SIGKDD Explorations Newsletter*, 14(2), 1–5.

Graumann, A., Houston, T., Lawrimore, J., Levinson, D., Lott, N., Mc-Cown, S., Stephens, S., and Wuertz, D. (2005). "Hurricane Katrina: A Climatological Perspective," 2005, NOAA Technical Report.

Haimes, Y. Y. (2004). *Risk Modeling, Assessment and Management*, New York: Wiley, 837–837.
ORourke, T. D. (2007). "Critical infrastructure interdependencies and resilience," *The Bridge*, 37(1), 22–29.
SIEMENS (2018). "Our Future Depends on Intelligent Infrastructures," Answers for Infrastructure and Cities [Online]. Available: www.siemens.com/intelligent-infrastructures.
The Federal Response to Hurricane Katrina (FRHK) (2006). "Lessons Learned," [Online]. Available: www.whitehouse.gov/infocus/hurricane/index.html.
The Subcommittee on Disaster Reduction (SDR) (2005). "Grand Challenges for Disaster Reduction," A Report of the Subcommittee on Disaster Reduction National Science and Technology Council [Online]. Available: www.sdr.gov.

8 Resilient Isolation-Structure Systems with Super-Large Displacement Friction Pendulum Bearings

Jinping Ou, Peisong Wu, and Xinchun Guan

CONTENTS

8.1 Introduction 123
8.2 Seismic Performance of Super-Large Displacement Friction Pendulum Bearing 125
    8.2.1 Mechanical Properties of Super-Large Displacement Friction Pendulum Bearings 125
    8.2.2 Seismic Performance of Isolated Structure with Super-Large Displacement Friction Pendulum Bearing 130
8.3 Seismic Performance of Prefabricated Structure Systems with Multiple Isolation Layers 135
8.4 Super-Large Displacement Translation Friction Pendulum Bearing 138
8.5 Conclusion 140
References 140

8.1 INTRODUCTION

The isolation method is classed as a passive control technique in the classification of structural vibration control. An isolation structure is a resilient structural system that is able to survive even in the worst conditions. There have been more than 1000 isolation structures built in China in the last 20 years. These projects demonstrated good isolation effects for different base isolation devices, including rubber bearing and friction pendulum bearing (Standard Institute of Chinese Construction, 2001; Zayas et al., 1990; Zayas et al., 1989). Many experimental tests have shown that Teflon or PTFE is suitable for the sliding surface of FPB because the friction coefficient is small and stable under long-term loads.

Ghobarah and Ali applied numerical analysis and experimental tests in order to evaluate isolation stiffness and frequency. They demonstrated that the real isolation frequency of FPB is larger than the theoretical frequency from a FPB model, and real horizontal stiffness is a larger than anticipated value (Ghobarah and Ali, 1988). A new model of FPB subjected to multiple components of excitation was presented by Mosqueda and Whittaker. They performed an investigation on the effects of vertical load. Their new model, with a prediction error smaller than 10%, is more applicable than an equivalent linear model with viscous damping (Mosqueda et al., 2004). The influence of isolation performance from vertical ground motions and the separation of sliding blocks and sliding surfaces was studied by Almazan and De la Llera (Almazan et al., 1998).

A challenge of vertical or 3D isolation system was also shown. There is no bearing with a large vertical carrying capacity and a small vertical stiffness. Vertical isolation bearing, which is possible and necessary when used in bridges or large span roofs, does not apply to heavy structures.

Another problem with an isolation structure is that tall isolation buildings will collapse with a large displacement of the base isolation layer or a large overturning moment (Yongfeng et al., 2011) of the superstructure under a super-strong earthquake. A tall isolation building designed to withstand an 8.0 earthquake may collapse with an 8.5 or 9.0 earthquake (Peisong et al., 2015). Increasing safety stock and robustness of an isolation structure, especially an isolation layer, is significant and relatively easy to realize because of the simple failure mode and certain weaknesses.

The seismic performance of an isolation layer, mainly its horizontal deformability, determines the security of isolation structures due to their certain failure modes. The deformability of isolation bearings have many restrictions, such as a small foundation space and limited horizontal stiffness. With reference to the theory of friction pendulum bearings, super-large displacement friction pendulum bearings are presented, as Figure 8.1 shows. Different from a classic friction pendulum bearing, located separately under each column, a super-large displacement friction pendulum bearing has one or several spherical shells of a large span and a large curvature radius as an integrated sliding isolation layer. The superstructure can sustain large horizontal displacement through relative sliding between the large spherical shell and sliding blocks under a nondeformable frame column.

The surface of the spherical shell and sliding block is Teflon, or some other low friction material. Each sliding block is connected by girders, columns, and other jointing elements as the superstructure. The resilience ability of SLDFPB is provided by the bearing reaction when the superstructure is sliding on the spherical shell. As Figure 8.1 shows, the motion of the superstructure in rotation and the deformation of the isolation layer can be huge. This horizontal deformation capability of SLDFPB can appease the displacement caused when it is subjected to super-strong earthquakes.

The horizontal stiffness of SLDFPB and its isolation frequency is determined by the design parameters of SLDFPB. The isolation frequency of SLDFPB is much smaller than the classic FPB. Prominent isolation effects also give rise to reducing the earthquake load and increasing the overturning capability. Each bearing reaction is pressure subjected to earthquakes, and the sliding block has no use for resistance to tension. The mass center of the superstructure and the stiffness center of the isolation layer are in the same vertical axis, so SLDFPB has good resistance to torsion.

A Novel Resilient Friction Pendulum Vibration Isolation System

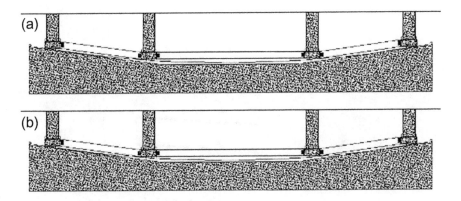

FIGURE 8.1 Super-large displacement friction pendulum bearings.

8.2 SEISMIC PERFORMANCE OF SUPER-LARGE DISPLACEMENT FRICTION PENDULUM BEARING

8.2.1 Mechanical Properties of Super-Large Displacement Friction Pendulum Bearings

The mechanical properties of a super-large displacement friction pendulum bearing are different from a classic friction pendulum bearing. To isolation devices of a sliding type, the horizontal component of the bearing reaction acts as a restoring force. In SLDFPB, the direction of the bearing reaction for each column is at the center of the spherical shell. To a classic pendulum bearing, each bearing reaction is straight up. The horizontal stiffness of SLDFPB and the classic pendulum bearing may have some differences.

The superstructure of a classic pendulum bearing moves along the orbit of the bearing and remains horizontal when subjected to earthquakes. With an isolation structure with SLDFPB, the superstructure rolls around along the orbit and each story leans toward the center shaft of the orbit. The seismic load is not only the seismic acceleration due to different directions of the seismic load and the motion of the superstructure. Stiffness between the isolation layer and each story of the superstructure is not zero because of the inclination of the superstructure.

The equation of motion for a frame structure in a superstructure of n stories and a SLDFPB can be easily calculated from the Lagrange Principle because the bearing reaction is unknown.

Figure 8.2 shows an isolation structure with a super-large displacement friction pendulum bearing. According to the Lagrange Principle, the equations of motion for the inclination angle of isolation layer θ and horizontal relative displacements u_i between each story of the superstructure and isolation layer are described by

$$\frac{d}{dt}\left(\frac{\partial T}{\partial \dot{\theta}}\right) - \frac{\partial T}{\partial \theta} + \frac{\partial V}{\partial \theta} = \frac{\partial (F_f + F_{gm})}{\partial \theta} \qquad (8.1)$$

$$\frac{d}{dt}\left(\frac{\partial T}{\partial \dot{u}_i}\right) - \frac{\partial T}{\partial u_i} + \frac{\partial V}{\partial u_i} = \frac{\partial (F_f + F_{gm})}{\partial u_i} \qquad (8.2)$$

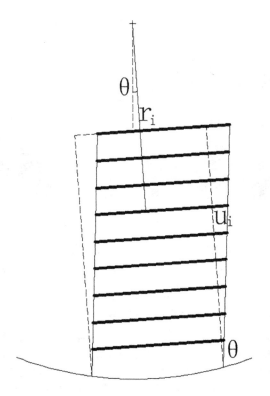

FIGURE 8.2 Isolation structure with super-large displacement friction pendulum bearing.

In which T = kinetic energy of the system, V = potential energy of the system, F_{gm} is load of the ground motion, F_f represents load of the friction force.

The results of Equations 8.1 and 8.2 are shown in Equations 8.3 and 8.4 when the frame structure is symmetrical.

$$\sum_{i=0}^{n} m_i r_i^2 \ddot{\theta} + \sum_{i=0}^{n} I_i \ddot{\theta} + \sum_{i=1}^{n} m_i \left(2 u_i \dot{u}_i \dot{\theta} + r_i \ddot{u}_i + u_i^2 \ddot{\theta} \right) + \sum_{i=0}^{n} m_i g r_i \sin\theta + \sum_{i=1}^{n} m_i g u_i \cos\theta$$
(8.3)
$$= m_0 \ddot{x}_g(t) r_0 \cos\theta + \sum_{i=1}^{n} m_i \ddot{x}_g(t) \left(r_i \cos\theta - u_i \sin\theta \right) + rf$$

$$m_i \left(\ddot{u}_i + r_i \ddot{\theta} \right) - m_i u_i \dot{\theta}^2 + m_i g \sin\theta + k_i u_i - k_i u_{i-1} + k_{i+1} u_i - k_{i+1} u_{i+1} = m_i \ddot{x}_g(t) \cos\theta \quad (8.4)$$

In which m_i is mass of *ith* story, r_i is distance between the circle center of the spherical orbit and center of the *ith* story, $I_i = \dfrac{1}{12} m_i L^2$ is the inertia moment of the ith story, L is the span of the frame, k_i is the horizontal stiffness of the *ith* story, r is the radius of spherical orbit, $i = 0$ represents the isolation layer. $k_{n+1} = 0$, $u_0 = 0$, which represents the unreal stiffness of the top story and relative displacement of the isolation layer and is convenient for describing the equation.

Equations 8.3 and 8.4 can be simplified when the angle is smaller than 5°. Assume $\sin\theta \approx \theta$, $\cos\theta \approx 1$.

$$\sum_{i=0}^{n} m_i r_i^2 \ddot{\theta} + \sum_{i=0}^{n} I_i \ddot{\theta} + \sum_{i=1}^{n} m_i \left(2 u_i \dot{u}_i \dot{\theta} + r_i \ddot{u}_i + u_i^2 \ddot{\theta}\right) + \sum_{i=0}^{n} m_i g r_i \theta + \sum_{i=1}^{n} m_i g u_i$$
$$= m_0 \ddot{x}_g(t) r_0 + \sum_{i=1}^{n} m_i \ddot{x}_g(t)(r_i - u_i \theta) + rf \qquad (8.5)$$

$$m_i\left(\ddot{u}_i + r_i\ddot{\theta}\right) - m_i u_i \dot{\theta}^2 + m_i g \theta + k_i u_i - k_i u_{i-1} + k_{i+1} u_i - k_{i+1} u_{i+1} = m_i \ddot{x}_g(t) \qquad (8.6)$$

Equations 8.5 and 8.6 are linearized because the results of high order items $m_i u_i \dot{u}_i \dot{\theta}$, $m_i u_i^2 \ddot{\theta}$, $m_i u_i \dot{\theta}^2$, and $m_i u_i \theta \ddot{x}_g(t)$ are much smaller than other items in the equation.

$$\left(\sum_{i=0}^{n} m_i r_i^2 + I_i\right) \ddot{\theta} + \sum_{i=1}^{n} m_i r_i \ddot{u}_i + \sum_{i=0}^{n} m_i g r_i \theta + \sum_{i=1}^{n} m_i g u_i \approx \sum_{i=0}^{n} m_i r_i \ddot{x}_g(t) + rf \qquad (8.7)$$

$$m_i\left(\ddot{u}_i + r_i\ddot{\theta}\right) + m_i g \theta + k_i u_i - k_i u_{i-1} + k_{i+1} u_i - k_{i+1} u_{i+1} = m_i \ddot{x}_g(t) \qquad (8.8)$$

Equations 8.9 and 8.10 describe the equations of motion in which $x_0 = r_0 \theta$ represents the horizontal displacement of the isolation layer.

$$\frac{\sum_{i=0}^{n} m_i r_i^2 + I_i}{\sum_{i=0}^{n} m_i r_i} m_0 \frac{\ddot{x}_0}{r_0} + m_0 \frac{\sum_{i=1}^{n} m_i r_i \ddot{u}_i}{\sum_{i=0}^{n} m_i r_i} + m_0 g \frac{x_0}{r_0} + m_0 \frac{\sum_{i=1}^{n} m_i g u_i}{\sum_{i=0}^{n} m_i r_i} - \frac{m_0 r}{\sum_{i=0}^{n} m_i r_i} f = m_0 \ddot{x}_g(t)$$
$$(8.9)$$

$$m_i \left(\ddot{u}_i + r_i \frac{\ddot{x}_0}{r_0}\right) + m_i g \frac{x_0}{r_0} + k_i u_i - k_i u_{i-1} + k_{i+1} u_i - k_{i+1} u_{i+1} = m_i \ddot{x}_g(t) \qquad (8.10)$$

In Equation 8.9, analogous to a classic friction pendulum bearing, the equivalent radius of SLDFPB r_e is defined by the relation between the isolation frequency and the radius of the isolation bearing. When the mass of each story is considered as the same value, the equivalent radius is described by

$$r_e = \frac{\sum_{i=0}^{n} m_i r_i^2 + I_i}{\sum_{i=0}^{n} m_i r_i} \approx r_0 - \frac{1}{2} nh + \frac{L^2 + (nh)^2 + 2nh^2}{12\left(r_0 - \frac{1}{2}nh\right)} \qquad (8.11)$$

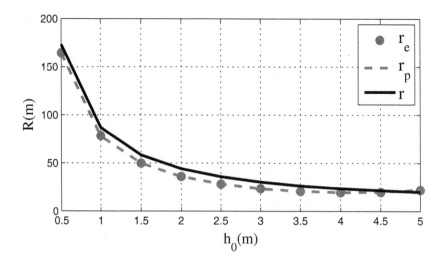

FIGURE 8.3 Equivalent radius of a frame structure with six stories.

Figure 8.3 shows the equivalent radius of a frame structure with six stories whose story height $h = 3m$, framing span $L = 26.3m$, and height of the isolation layer $h0$ is from 0.5 to 5 meters. $r_p = \dfrac{g}{\omega_1^2}$ represents the equivalent radius calculated from the nature period of the isolation system.

Figure 8.3 shows that the result of equivalent radius r_e is consistent with the equivalent radius r_p and it is unequal to the radius of the spherical orbit. The nature period of the isolation system can be evaluated accurately by equivalent radius r_e not r.

Figure 8.4 is the equivalent radius of different isolation layers used in frame structures with different stories. Both the radius of the spherical orbit and the equivalent

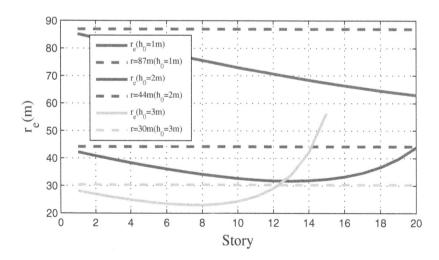

FIGURE 8.4 Equivalent radius of different frame structures.

radius decrease with the height of the isolation layer h_0 in the same situation. To a certain SLDFPB, equivalent radius r_e is determined by the height of the superstructure. r_e is smaller than r when the height of the superstructure is smaller than the radius of the spherical orbit. The smallest equivalent radius appears when the height of the superstructure is a little larger than the radius of the spherical orbit. It is possible for the equivalent radius to exceed the radius of the spherical shell when the height of the structure is larger than the radius.

Based on the Coulomb Friction Principle, $f = -\text{sgn}(\dot{x}_0) \sum \mu N$, items of friction force in an equation of motion are related to the algebraic sum of the bearing reaction when the friction coefficient and curvature radius is constant everywhere. The result of the bearing reaction is hard to obtain from the Lagrange Principle.

In a single-span multi-story frame structure, following D'Alembert's Principle, the horizontal and vertical resultant force is equal to zero when the superstructure is stationary at a position of equilibrium and the friction force is zero. The equation of equilibrium is described by

$$\begin{bmatrix} \cos\alpha_1' & \cos\alpha_2' \\ \sin\alpha_1' & \sin\alpha_2' \end{bmatrix} \begin{bmatrix} N_1 \\ N_2 \end{bmatrix} = \begin{bmatrix} \sum m_i g + \sum m_i r_i \sin(\theta + \alpha_i) \ddot{\theta} \\ \sum m_i \ddot{x}_g - \sum m_i r_i \cos(\theta + \alpha_i) \ddot{\theta} \end{bmatrix} \quad (8.12)$$

$$\begin{bmatrix} N_1 \\ N_2 \end{bmatrix} \approx \frac{m}{\sin(2\alpha')} \begin{bmatrix} \theta\left[\ddot{x}_g \sin\alpha' + g\cos\alpha'(1-\lambda)\right] + \left[g\sin\alpha' - \ddot{x}_g \cos\alpha'(1-\lambda)\right] \\ \theta\left[\ddot{x}_g \sin\alpha' - g\cos\alpha'(1-\lambda)\right] + \left[g\sin\alpha' + \ddot{x}_g \cos\alpha'(1-\lambda)\right] \end{bmatrix} \quad (8.13)$$

$$\lambda = \frac{r_0 - \frac{1}{2}nh}{r_e} \quad (8.14)$$

In which $\alpha' = -\alpha_1' = \alpha_2'$ represents the angle between the vertical direction and the direction of the sliding block to the center of the spherical orbit; N_1, N_2 represents the bearing reactions of the columns; α_i represents the angle between the vertical direction and the direction from the edge of the ith story to the center of the spherical shell, λ is a coefficient for calculating bearing reactions. The algebraic sum of the bearing reaction is described by

$$N_1 + N_2 \approx \sum_{i=0}^{n} m_i g \frac{r_0}{r} \quad (8.15)$$

The algebraic sum of the bearing reaction to movement changes a little when the friction coefficient is small. This conclusion also applies to statically indeterminate structures such as multi-span frame structures. The equation of motion can be described by replacing the bearing reaction with the gravity of the superstructure because r_0 and r have very few differences. Plug the result into Equation 8.9, and the equation of motion for the isolation layer is described by

$$\frac{\sum_{i=0}^{n} m_i r_i^2 + I_i}{\sum_{i=0}^{n} m_i r_i} m_0 \frac{\ddot{x}_0}{r_0} + m_0 \frac{\sum_{i=1}^{n} m_i r_i \ddot{u}_i}{\sum_{i=0}^{n} m_i r_i} + m_0 g \frac{x_0}{r_0} + m_0 \frac{\sum_{i=1}^{n} m_i g u_i}{\sum_{i=0}^{n} m_i r_i} + \frac{\sum_{i=0}^{n} m_i r_0}{\sum_{i=0}^{n} m_i r_i} \mu m_0 g \operatorname{sgn}(\dot{x}_0) = m_0 \ddot{x}_g(t)$$

(8.16)

Equation 8.16 can be rewritten in a similar form to the equation of motion for a friction pendulum bearing.

$$\frac{r_e}{r} m_0 \ddot{x}_0 + \mu_e m_0 g \operatorname{sgn}(\dot{x}_0) + \frac{m_0 g}{r} x_0 = m_0 \ddot{x}_g \qquad (8.17)$$

In which,

$$r_e = \frac{\sum_{i=0}^{n} m_i r_i^2 + I_i}{\sum_{i=0}^{n} m_i r_i} < r \qquad (8.18)$$

$$\mu_e = \frac{\sum_{i=0}^{n} m_i r_0}{\sum_{i=0}^{n} m_i r_i} \mu > \mu \qquad (8.19)$$

r_e and μ_e is defined by the same relation of the frequency and damping ratio to the radius and friction coefficient of the isolation bearing. It's a linear system because the stiffness and damping of the system are constant. The equivalent radius is smaller than the radius of the spherical shell and the equivalent friction coefficient is larger than the friction coefficient of the spherical shell. The equation of motion shows that SLDFPB and classic FPB have the same ratio between restoring force and earthquake load; SLDFPB has a larger ratio between friction force and earthquake load, but a lower ratio between inertia force and earthquake load.

Because the number, size, and distance of the column have nothing to do with the equation of motion, SLDFPB can keep its mechanical property even if several columns separate from the spherical shell caused by an enormous deformation of the isolation layer or bad contact with the sliding block. SLDFPB has a stable performance of mechanical property.

8.2.2 Seismic Performance of Isolated Structure with Super-Large Displacement Friction Pendulum Bearing

The seismic performance of an isolation structure with different parameters of SLDFPB subjected to rare and super-rare earthquakes was studied. A frame

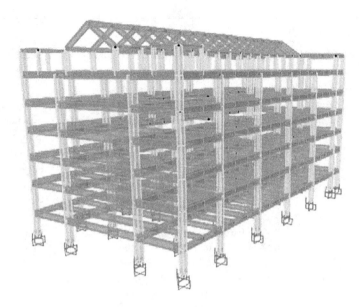

FIGURE 8.5 Frame structure with six stories.

structure of six stories whose natural period is 0.77 s represents a low-rise structure, as Figure 8.5 shows. The height, length, width, and damping ratio of this structure is 27.3 meters, 48.1 meters, 26.3 meters, and 0.05, respectively.

The main parameters of SLDFPB are its height and friction coefficient. Height, for example, 1 meter, 2 meters, 3 meters, determines the radius of the spherical shell and the natural frequency of the bearing; the friction coefficient is determined by the materials of interface. Generally, the friction coefficient of a liquid or graphite lubricant can reach 0.01; the friction coefficient of steel and concrete is nearer to 0.02; the friction coefficient of a classic friction pendulum bearing is usually between 0.04 and 0.1.

Five natural earthquakes and two artificial earthquakes were used in a time history analysis. Each ground motion was calculated at a different peak ground acceleration which represented a different intensity of earthquake.

The mean value of the response when subjected to seven ground motions is shown in Tables 8.1 and 8.2.

Tables 8.1 and 8.2 show the main response of the isolation layer and the superstructure separately. This frame structure had some plastic damage when subjected to an earthquake whose PGA was 0.2 g. SLDFPB can survive when subjected to an earthquake whose PGA is 0.62 g. The horizontal deformation of the isolation layer is sustainable, and the superstructure has a little plastic damage. The response of the inter-story draft ratio shows that SLDFPB has a great isolation effect because of its large vertical carrying capacity and small horizontal stiffness. When benefiting from a large space in the isolation layer, SLDFPB can have a larger bearing radius and a larger equivalent radius. Its horizontal stiffness could be small enough for structures to survive when subjected to super-strong earthquakes.

SLDFPB with a smaller height of isolation layer h_0 has a larger equivalent radius r_e, a larger isolation period, and a better isolation effect.

TABLE 8.1
Response of Isolation Layer Subjected to Rare and Super-Strong Earthquakes

Height of Isolation Bearing	Friction Coefficient	Maximum Disp. of Isolation Layer (m)		Residual Disp. of Isolation Layer (m)		Maximum Reaction Force		Minimum Reaction Force	
PGA		0.4 g	0.62 g	0.4 g	0.62 g	0.4 g	0.62 g	0.4 g	0.62 g
1 m	0.01	0.48	0.78	0.090	0.114	103.7%	105.8%	96.2%	94.0%
	0.02	0.38	0.70	0.108	0.156	103.7%	105.8%	96.2%	94.1%
	0.04	0.34	0.68	0.122	0.207	103.7%	105.8%	96.2%	94.1%
	0.1	0.19	0.33	0.060	0.134	103.8%	105.8%	96.2%	94.1%
2 m	0.01	0.48	0.87	0.061	0.096	108.5%	113.2%	91.3%	86.3%
	0.02	0.38	0.67	0.080	0.077	108.5%	113.2%	91.4%	86.5%
	0.04	0.37	0.67	0.114	0.153	108.6%	113.2%	91.4%	86.6%
	0.1	0.16	0.27	0.067	0.127	108.6%	113.2%	91.4%	86.6%
3 m	0.01	0.52	0.90	0.040	0.069	113.6%	121.0%	86.1%	78.3%
	0.02	0.42	0.74	0.065	0.058	113.6%	121.0%	86.2%	78.5%
	0.04	0.23	0.55	0.060	0.109	113.6%	121.0%	86.4%	78.7%
	0.1	0.18	0.27	0.054	0.079	113.6%	121.1%	86.4%	78.8%

TABLE 8.2
Response of Superstructure Subjected to Rare and Super-Strong Earthquakes

Height of Isolation Bearing	Friction Coefficient	Inter-Story Draft Ratio		Tangential of Maximum Inclination Angle		Maximum Vertical Displacement of Upper Structure (m)		Maximum Vertical Acceleration (m/s²)	
PGA		0.4 g	0.62 g	0.4 g	0.62 g	0.4 g	0.62 g	0.4 g	0.62 g
1 m	0.01	1/889	1/639	0.003	0.006	0.043	0.073	0.009	0.024
	0.02	1/623	1/513	0.003	0.005	0.041	0.065	0.008	0.021
	0.04	1/392	1/337	0.002	0.005	0.032	0.063	0.007	0.020
	0.1	1/252	1/194	0.000	0.002	0.013	0.031	0.004	0.013
2 m	0.01	1/430	1/282	0.007	0.011	0.086	0.150	0.019	0.053
	0.02	1/380	1/270	0.006	0.009	0.075	0.125	0.016	0.043
	0.04	1/284	1/224	0.005	0.008	0.060	0.112	0.014	0.037
	0.1	1/188	1/155	0.002	0.005	0.025	0.061	0.009	0.025
3 m	0.01	1/249	1/159	0.010	0.017	0.131	0.226	0.033	0.086
	0.02	1/249	1/163	0.009	0.015	0.112	0.193	0.027	0.075
	0.04	1/210	1/154	0.006	0.012	0.073	0.155	0.018	0.056
	0.1	1/144	1/114	0.002	0.006	0.028	0.073	0.009	0.036

A Novel Resilient Friction Pendulum Vibration Isolation System

The inclination angle of superstructure θ is tiny even when subjected to super-strong earthquakes, so there is little influence of the rotation motion on personnel and equipment.

The algebraic sum of the bearing reaction changes a little when it is subjected to super-strong earthquakes; the friction coefficient has little influence on each bearing reaction.

The maximum vertical displacement of the superstructure appears at the edge of the frame. This response is mainly caused by the vertical displacement of the isolation layer at the edge of the frame and the level of tilt.

The maximum vertical acceleration of the superstructure is also not big. There is no strong sense of weightlessness or heaviness when subjected to super-strong earthquakes.

Figure 8.6 shows the inter-story draft ratio of an isolation structure whose height of isolation bearing h_0 is 1 meter and 2 meters when subjected to earthquakes whose PGA is 0.4 g. It has an excellent isolation effect when the friction coefficient is small. There is a similar isolation effect when the friction coefficient between 0.01 to 0.02. SLDFPB with a relatively small friction coefficient can have a sufficient isolation capability.

Figure 8.7 shows the maximum displacement of an isolation layer with $h_0 = 1m$. With the increase of the friction coefficient, the deformation of the isolation layer reduces. The maximum displacement of SLDFPB with a small friction coefficient is similar because the response is close to the horizontal displacement of the ground as

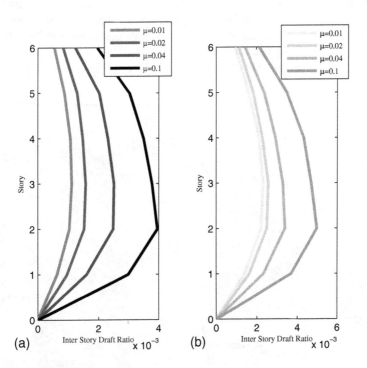

FIGURE 8.6 Inter-story draft ratio of isolation structure, h0=2m.

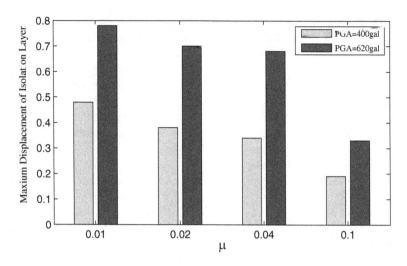

FIGURE 8.7 Maximum displacement of isolation layer of isolation structure with h0=1m.

the result of a frictionless surface. The maximum displacement of SLDFPB when subjected to super-strong earthquakes is large but much smaller than the deformation capability of SLDFPB. The structure will not collapse due to the deformation of SLDFPB.

Figure 8.8 shows the residual displacement of an isolation layer with PGA = 400 gal. Residual displacement is a random variable determined by the status of the bearing during the end of the ground motion when subjected to different ground motions. It should be smaller than both the maximum displacement of SLDFPB and the maximum theoretical residual displacement. Both of them are determined by the friction coefficient and the equivalent radius. The maximum theoretical residual

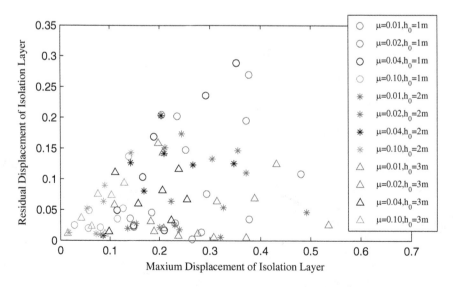

FIGURE 8.8 Residual displacement of isolation layer with PGA=400gal.

displacement is the position where the restoring force is equal to the friction force. SLDFPB with a large height of isolation layer h_0 and small equivalent radius r_e has a small maximum displacement and small maximum theoretical residual displacement, so the residual displacement is also probably small.

SLDFPB with a small friction coefficient and large maximum displacement will not restrict residual displacement; both maximum theoretical residual displacement and residual displacement reduce with a decrease of the friction coefficient. On the contrary, to a SLDFPB with a large friction coefficient, residual displacement is determined by a small maximum displacement; both maximum displacement and residual displacement increase with a decrease of the friction coefficient. The result indicates that considering reset ability separately, the small height of isolation layer h_0 is recommended and the suggested range of the friction coefficient between 0.01 and 0.1.

Figure 8.9 shows the maximum and minimum reaction force of SLDFPB with different heights of isolation layer h_0. Overturning safety is ensured when all bearing reactions are under compressionChanges in the reaction force are almost unrelated to the friction coefficient. The reaction force only changes a little when the equivalent radius is very large. At the same time, the safety stock of its vertical carry capability is easily satisfied.

Considering the seismic performance of the isolation structure, a small friction coefficient will lead to the small maximum residual displacement of the isolation bearing and a good isolation effect; large equivalent radius is good for keeping away from the frequency of ground motion.

8.3 SEISMIC PERFORMANCE OF PREFABRICATED STRUCTURE SYSTEMS WITH MULTIPLE ISOLATION LAYERS

Referring to the theory of tuned mass dampers, a super-large displacement friction pendulum bearing can be applied to combine isolation with an energy-dissipation

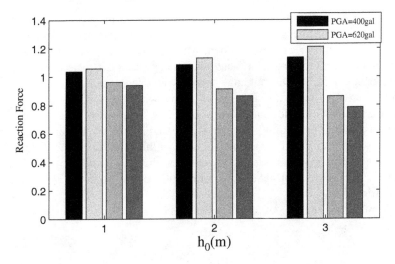

FIGURE 8.9 Maximum and minimum reaction force.

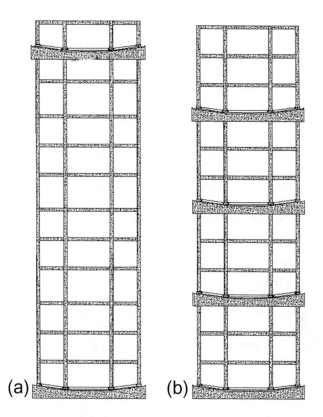

FIGURE 8.10 Schematic diagram of prefabricated building system with multiple isolated layers.

technique. A new system that has two isolation layers in the top and base, respectively, was studied, as Figure 8.10(a) shows. The main advantage of a base isolation layer is to ensure an elastic system when subjected to rare earthquakes. The top isolation layer working as a tuned mass damper can dissipate the earthquake. Displacement of the TMD is not huge because of the large mass.

Figure 8.10(b) shows an isolation structure with multi isolation layers. The high rise structure divides into several modules, each with some stories. Each module is a whole isolation structure with SLDFPB. Each module can remain resilient when subjected to rare earthquakes, so this structure system has a good resilience ability. The complicated structure with many degrees of freedom and modes of vibration simplifies to several degrees of freedom and evident modes of vibration.

Optimized parameters of an isolation system with two isolation layers are designed using

$$f_\omega = \frac{1}{1+\mu_m} \tag{8.20}$$

$$\zeta_c = \sqrt{\frac{3\mu_m}{8(1+\mu_m)}} \tag{8.21}$$

In which f_ω is the frequency ratio of the top isolation layer, μ_m is the mass ratio of the top isolation layer, ζ_c is the damping ratio of the top isolation layer.

According to the same energy dissipation of one hysteretic loop, the relationship between the equivalent friction coefficient and the damping ratio is described by

$$\zeta_c = \frac{E_D}{2\pi k d_{max}^2} = \frac{\mu_e r_e}{2\pi d_{max}} \tag{8.22}$$

The optimized size of the top isolation layer and the friction coefficient of the top spherical orbit can be determined using Equations 8.10, 8.11, 8.12, and 8.15.

Super-large displacement friction pendulum bearings also apply to an isolation structure with multiple isolation layers. The seismic response of this system and two examples of an isolation structure with two isolation layers are compared and shown in Table 8.4. Figure 8.11 is a superstructure with 20 stories whose natural period is 2.1 s. Parameters of SLDFPB in the examples are shown in Table 8.3.

The seismic performance of isolation structures with two or multiple isolation layers are indicated in Table 8.4. The maximum displacement of the base isolation

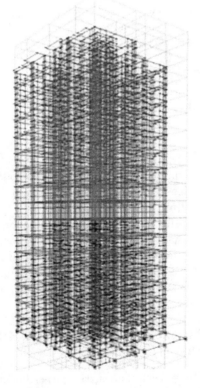

FIGURE 8.11 Superstructure with 20 stories in the examples.

TABLE 8.3
Parameters of SLDFPB in the Examples

	System A with Two Isolation Layers	System B with Two Isolation Layers	System C with Multiple Isolation Layers
Location (num. of story)	20	18	1,7,14
Height (m)	1.237	0.756	1
Friction coefficient	0.0116	0.0127	0.01

TABLE 8.4
Seismic Response of Three Isolation Systems

	System A with Two Isolation Layers		System B with Two Isolation Layers		System C with Multiple Isolation Layers		System D with One Isolation Layer	
PGA	0.4 g	0.6 g	0.4 g	0.6 g	0.4 g	0.6 g	0.4 g	0.6 g
Max Disp. of Base Isolation Layer (m)	0.507	0.953	0.507	0.937	0.530	1.029	0.613	1.124
Residual Disp. of Base Isolation Layer (m)	0.116	0.144	0.122	0.157	0.093	0.163	0.091	0.213
Max Disp. of Top Isolation Layer (m)	0.359	0.569	0.263	0.458	0.008	0.014	–	–
Residual Disp. of Top Isolation Layer (m)	0.038	0.087	0.057	0.078	0.001	0.003	–	–
Inter-story Draft Ratio	1/303	1/214	1/545	1/344	1/294	1/203	1/325	1/212
Max Disp. of Mid Isolation Layer (m)	–	–	–	–	0.010	0.012	–	–
Residual Disp. of Mid Isolation Layer (m)	–	–	–	–	0.002	0.002	–	–

layer in system A and B is similar, so they have similar isolation effects. The seismic performance of the isolation systems is determined by the displacement of the top isolation layer and the control effect.

Only system B has a better control effect than isolation-based structures with SLDFPB. A large ratio of mass and optimized parameters contribute to a good control effect. The design of isolation layers has a large influence on seismic performance. One group of inter-story draft ratios is shown in Figure 8.12. The inter-story draft ratio of each story in system B is smaller and closer, and the seismic performance of each story is fully exerted.

8.4 SUPER-LARGE DISPLACEMENT TRANSLATION FRICTION PENDULUM BEARING

Another kind of super-large displacement friction pendulum bearing was studied, as Figure 8.13 shows. There are a group of spherical shells with a large radius and

A Novel Resilient Friction Pendulum Vibration Isolation System

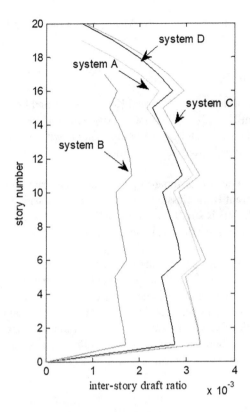

FIGURE 8.12 Inter-story draft ratio of four system.

span in the isolation layer. Several sliding frames used as basements or mechanical floors act as a sliding block for the friction pendulum bearing. This bearing and FPB have similar mechanical properties because there's no rotation. Subjected to earthquakes, the superstructure will move along its arc and keep vertical. All these super-large displacement friction pendulum bearings are studied by Key Lab of Structures

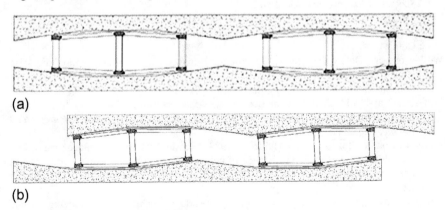

FIGURE 8.13 Super-large displacement translation friction pendulum bearing.

Dynamic Behaviour and Control of the Ministry of Education, Harbin Institute of Technology.

8.5 CONCLUSION

This research is supported by the National Key Research and Development Program of China with Grant No. 2017YFC0703603. Two kinds of super-large displacement friction pendulum bearings and two isolation systems with multiple isolation layers are introduced. The conclusions are as follows:

1. Mechanical properties of super-large displacement friction pendulum bearings are different from classic friction pendulum bearings. The equivalent radius of SLDFPB is smaller than the radius of the spherical shell, and the equivalent friction coefficient is larger than the friction coefficient of the contact surface.
2. SLDFPB has a large vertical carrying capacity, small horizontal stiffness, and a great isolation effect. The friction coefficient has a large influence on seismic performance and resilience ability.
3. Two resilient-isolated structural systems that combined isolation with energy-dissipation techniques were studied. A large ratio of mass and optimized parameters contributed to a good isolation effect. The seismic performance of an isolation system is determined by the design of the isolation layers.

REFERENCES

Almazan J. L., De La Llera J. C., and Inaudi J. A. (1998) "Modelling aspects of structures isolated with the frictional pendulum system," *Earthquake Engineering & Structural Dynamics*, 27(8): 845–867.

Ghobarah A. and Ali H. M. (1988) "Seismic performance of highway bridges," *Engineering Structures*, 10(3): 157–166.

Mosqueda G., Whittaker A. S., and Fenves G. L. (2004) "Characterization and modeling of friction pendulum bearings subjected to multiple components of excitation," *Journal of Structural Engineering*, 130(3): 433–442.

"Technical specification for seismic-isolation with laminated rubber bearing isolators" (2001). Technical specification Enacted by the Standard Institute of Chinese Construction, China.

Wu Peisong and Ou Jinping (2015) "Seismic performance analysis and design of high-rise structures under super-strong earthquake," *6th International Conference on Advances in Experimental Structural Engineering*, UIUC, USA.

Yongfeng D. and Hui L. (2011) "Numerical analysis on overturning resistant property of seismic isolated building subject to Bi-directional earthquake excitation," *Computer Aided Engineering*, 20(1): 42–46.

Zayas V., Low S., Bozzo L., and Mahin S. (1989) "Feasibility and performance studies on improving the earthquake resistance of new and existing buildings using the friction pendulum system," Technical Report UBC /EERC-89 /09, University of California at Berkeley.

Zayas V., Low S., and Mahin S. (1990) "A simple pendulum technique for achieving seismic isolation," *Earthquake Spectra*.

9 Real-Time City-Scale Time-History Analysis and Its Application in Resilience-Oriented Earthquake Emergency Response

Xinzheng Lu, Qingle Cheng, Zhen Xu, Yongjia Xu, and Chujin Sun

CONTENTS

9.1	Introduction	142
9.2	Real-Time City-Scale Nonlinear Time-History Analysis	143
	9.2.1 Framework	143
	9.2.2 Real-Time Recorded Ground Motions	143
	9.2.3 Building Inventory Database	143
	9.2.4 City-Scale Nonlinear Time-History Analysis	145
	9.2.4.1 Parameter Determination Method for Buildings in China	145
	9.2.4.2 Parameter Determination for Backbone Curve Based on the HAZUS Data	147
	9.2.5 Regional Seismic Loss Prediction and Resilience Assessment	149
	9.2.5.1 Regional Seismic Loss Prediction Using Conventional Method	149
	9.2.5.2 Regional Resilience Assessment Using FEMA P-58	151
	9.2.6 High-Performance Computing for Post-Earthquake Emergency Response	153
9.3	Applications in Earthquake Emergency Response	153
	9.3.1 Overview of the Applications	153
	9.3.2 2017 M7.0 Jiuzhaigou Earthquake	155
	9.3.3 2018 Mw 7.0 Anchorage Earthquake	157
9.4	Conclusions	157
Acknowledgments		159
References		159

9.1 INTRODUCTION

Earthquakes cause severe damage and economic loss to urban areas, and the resilience of cities to this damage has received worldwide attention. An accurate and rapid assessment of seismic damage, economic loss, and post-event repair time can provide an important reference for emergency rescue and post-earthquake recovery. Therefore, it is of great importance to the enhancement of community resilience.

The experience of several major earthquakes in recent years indicates that the assessment of building damage in an earthquake-stricken area needs to be further improved. After an earthquake, communication in the disaster area is delayed, the disaster site is usually chaotic, and there are not enough professionals to quickly evaluate building safety. Furthermore, rumors and fake information on the internet may interfere with accurate seismic damage assessment. Therefore, it is necessary to propose a scientific, objective, and timely method for earthquake loss assessment.

To date, the near-real-time earthquake loss estimation tools primarily include Prompt Assessment of Global Earthquakes for Response (PAGER), Global Disaster Alert and Coordination System (GDACS), USGS-ShakeCast, the Istanbul Earthquake Rapid Response System, and the Rapid Response and Disaster Management System in Yokohama, Japan, etc. (Erdik et al., 2011). These seismic loss estimation systems generally comprise three parts: the ground motion intensity measure (IM), building inventory and fragility, and direct economic losses and casualties. The ground motion IM can be obtained directly from the real-time monitoring data of a seismic network or calculated using ground motion prediction equations (GMPE). The building inventory data can be determined using either a detailed building database or macroscopic statistical data. The seismic damage to buildings can be predicted using the damage probability matrix (DPM) method or the capacity spectrum method. Economic loss and casualties are generally calculated using empirical models.

However, the main problems existing in these systems are as follows: (a) the dynamic characteristics of ground motion are not comprehensively considered; (b) the DPM method is difficult to apply in areas where historical earthquake data are lacking or in quickly developing areas where there are large differences between the inventories of current and historical buildings; (c) the capacity spectrum method cannot easily represent the concentration of damage to different stories or the time-domain properties of ground motions (e.g., the velocity impulse of ground motions); (d) the earthquake loss prediction method relies on historical seismic damage data, and the repair time cannot be provided in these systems.

Consequently, this work proposes a real-time city-scale time-history analysis method for post-earthquake damage assessment. The actual ground motion records obtained from seismic stations are input into the building models of the earthquake-stricken area, and the nonlinear time-history analysis of these models is subsequently performed. The seismic damage, economic loss, and repair time of the target region subjected to this earthquake are evaluated according to the analysis result. A program, named "Real-time Earthquake Damage Assessment using City-scale Time-history analysis" (or "RED-ACT" for short), is developed. The application and the advantages of the proposed method are demonstrated through actual earthquake events.

9.2 REAL-TIME CITY-SCALE NONLINEAR TIME-HISTORY ANALYSIS

9.2.1 Framework

The proposed framework to conduct the real-time city-scale time-history analysis and loss assessment is illustrated in Figure 9.1. The corresponding procedures are as follows:

(1) Obtaining the real-time ground motion records from the seismic stations
(2) Establishing the building inventory database for the target region
(3) Conducting the city-scale nonlinear time-history analysis to predict the seismic damage of the target region
(4) Performing the regional seismic loss prediction to assess the seismic economic loss and repair time of the target region

9.2.2 Real-Time Recorded Ground Motions

The ground motion records can fully describe the features of the ground motions with no information loss. The densely distributed seismic stations and communication networks make it possible to obtain real-time ground motion records. After an earthquake, the ground motion record near the epicenter can be quickly obtained through the seismic stations and communication network, and information such as the station's latitude, longitude, and recording time can be collected simultaneously. With the development of monitoring and data-transforming technology, the densely distributed strong motion network will cover more regions, and the ground motion data will be easier to access in a timely manner after an earthquake.

9.2.3 Building Inventory Database

Based on the Sixth National Population Census (Population Census Office under the State Council & Department of Population and Employment Statistics of National Bureau of Statistics, 2012), this work constructs a virtual building inventory database of cities in the mainland of China. Specifically, according to the Sixth National Population Census, the number of buildings in the target region classified by the number of stories, structural type, and the year they were built can be obtained. The buildings are divided into 33 categories according to the number of stories, structural type, and year built, and the proportions of the 33 building types can be determined by solving the indefinite equations that describe this problem. The building inventory database of each region can then be established to serve the subsequent seismic damage prediction. Note that if the statistical data of each building can be obtained for the target region, then these data can be directly used to establish the analysis model. In addition to the cities in the mainland of China, other building inventory databases for other regions (e.g., Japan and the United States) are under construction. As a result, the proposed method can be further applied to different regions if the ground motions and building inventory are available.

144 Resilience of Critical Infrastructure Systems

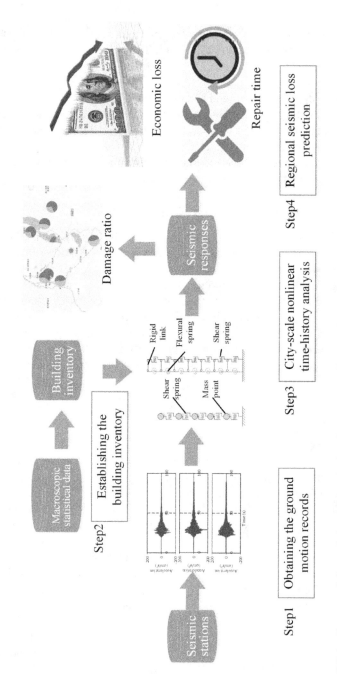

FIGURE 9.1 Framework for real-time city-scale time-history analysis and loss assessment.

9.2.4 City-Scale Nonlinear Time-History Analysis

The city-scale nonlinear time-history analysis is used to perform the seismic damage simulation for the target region (Lu & Guan, 2017). During the analysis, the buildings are divided into two types: ordinary multi-story buildings and ordinary tall buildings. In general, multi-story buildings often exhibit shear deformation modes under earthquakes, whereas tall buildings will deform in flexural-shear modes. Thus, the multiple-degree-of-freedom (MDOF) shear model will be used for the multi-story buildings (Figure 9.2A), and the MDOF flexural-shear model will be applied to tall buildings (Figure 9.2B). For the MDOF model, the masses of the buildings are concentrated on their corresponding stories, and the nonlinear behavior of the structure is represented by the nonlinear inter-story force–displacement relationships (Figure 9.2C). The tri-linear backbone curves recommended in the HAZUS report (FEMA, 2012a) are employed to model the inter-story force–displacement relationships. Note that the parameter determination of the inter-story force–displacement relationships is critical for the rationality and accuracy of the simulation results, considering the limited available information for buildings on a regional scale. The parameter determination methods for buildings in China and the United States proposed by the authors (Lu & Guan, 2017) are adopted in the city-scale nonlinear time-history analysis. For buildings in other countries, the parameter determination procedure can refer to these two methods.

9.2.4.1 Parameter Determination Method for Buildings in China

The parameter determination method for buildings in China is based on the design codes and statistics of extensive experimental and analytical results (Xiong et al., 2016, 2017). The elastic parameters of a building can be represented by the inter-story shear stiffness, k_0, and the mass, m, of each story. Equations 9.1 and 9.2 show the global stiffness [K] and mass matrices [M] of a structure with a uniform stiffness and mass along the height (Lu et al., 2014).

$$[K] = k_0 \begin{bmatrix} 2 & -1 & & & \\ -1 & 2 & -1 & & \\ & -1 & \ddots & \ddots & \\ & & \ddots & 2 & -1 \\ & & & -1 & 1 \end{bmatrix} = k_0 [A] \qquad (9.1)$$

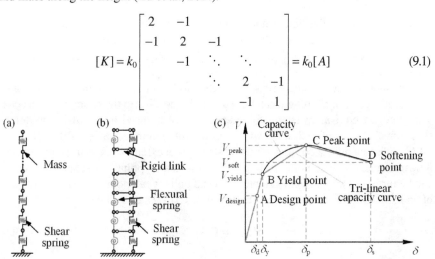

FIGURE 9.2 (A) MDOF shear model; (B) MDOF flexural-shear model; and (C) Trilinear backbone curve adopted in the MDOF model (Xiong et al. 2016, Xiong et al. 2017).

$$[M] = m \begin{bmatrix} 1 & & & & \\ & 1 & & & \\ & & 1 & & \\ & & & \ddots & \\ & & & & 1 \end{bmatrix} = m[I] \qquad (9.2)$$

The mass of each story, m, in Equation 9.2 can be determined based on the area of each story, A_1, and the mass per unit area, m_1 (Equation 9.3) (Sobhaninejad et al., 2011); m_1 can be estimated according to the occupancy of each story.

$$m = m_1 A_1 \qquad (9.3)$$

The relationship among the stiffness, mass, and first vibration period, T_1, can be expressed using Equation 9.4.

$$k_0 = m\omega_1^2 \left(\frac{[\Phi_1]^T [I][\Phi_1]}{[\Phi_1]^T [A][\Phi_1]} \right) = \frac{4\pi^2 m}{T_1^2} \left(\frac{[\Phi_1]^T [I][\Phi_1]}{[\Phi_1]^T [A][\Phi_1]} \right) \qquad (9.4)$$

where $[\Phi_1]$ is the first mode vector. Given the stiffness matrix $[K]$ and mass matrix $[M]$, $[\Phi_1]$ can be computed using a generalized eigenvalue analysis. As shown in Equation 9.4, m and T_1 are required to obtain k_0. The vibration periods of different types of structures can be estimated using empirical equations. For example, the fundamental period of a reinforced concrete (RC) frame can be calculated using the empirical equation (Equation 9.5) specified in the Chinese Code (GB50009-2012).

$$T_1 = 0.25 + 0.00053 H^2 / \sqrt[3]{B} \qquad (9.5)$$

where H and B are the height and width of an RC frame, respectively. For other types of structures (e.g., RC tall buildings, masonry structures, steel frames), the corresponding empirical vibration period can also be adapted to determine the interstory stiffness.

For engineering designed structures (e.g., RC frames and reinforced masonry structures), the design strength can be estimated according to the seismic design code. Therefore, an equivalent lateral force analysis can be used to calculate the design shear force, $V_{\text{design}, i}$, of each story, where i is the story number (ASCE, 2010; GB50011-2010). Subsequently, according to related statistics of extensive experimental and analytical results, the yield point, peak point, and softening point on the backbone curve can be further obtained. For example, the yield point, peak point, and softening point of an RC frame can be determined by Equations 9.6–9.8 as follows:

$$V_{\text{yield}, i} = \Omega_1 V_{\text{design}, i} \qquad (9.6)$$

$$V_{\text{peak}, i} = \Omega_2 V_{\text{design}, i} \qquad (9.7)$$

$$V_{\text{ultimate}, i} = V_{\text{peak}, i} \qquad (9.8)$$

where $V_{\text{yield}, i}$, $V_{\text{peak}, i}$, and $V_{\text{ultimate}, i}$ are the yield strength, peak strength, and ultimate strength, respectively. Ω_1 is the yield overstrength ratio of RC frames, which is determined according to the partial factor of the steel reinforcement (GB50010-2010). Ω_2 is the peak overstrength ratio, which is determined by the statistics of 155 pushover results of RC frames designed following the Chinese seismic design code. The deformation parameters of RC frames, including the yield, peak, and ultimate deformations, can be determined using the same procedure.

Non-engineered buildings (e.g., unreinforced masonry buildings and adobe buildings) lack a design strength as the reference strength to establish the backbone curve in Figure 9.2C. Consequently, the statistical strengths of different types of non-engineered buildings obtained from the literature are used to establish the backbone curve. For example, Xiong et al. (2016) proposed using the statistical results of peak strength per unit area of 1000 unreinforced masonry structures in China as the reference strength. The other parameters on the backbone curve of unreinforced masonry buildings are determined based on the statistical results of 97 unreinforced masonry wall experiments.

9.2.4.2 Parameter Determination for Backbone Curve Based on the HAZUS Data

Based on the HAZUS database, Lu et al. (2014) also proposed a modeling approach by which all parameters of the MDOF shear model can be determined from the basic building information (i.e., number of stories, height, year built, structural type, floor area, and occupancy). Specifically, the first vibration period of the building can be determined by the typical buildings presented in Tables 5.5 and 5.7 of FEMA (2012a). The elastic parameters of a building, including the inter-story shear stiffness, k_0, and the mass, m, are determined using Equations 9.1–9.4. Subsequently, the inter-story backbone curve parameters of story i in Figure 9.3 are determined as follows:

$$k_{0,i} = \lambda \frac{4\pi^2 m}{T_1^2} \quad (9.9)$$

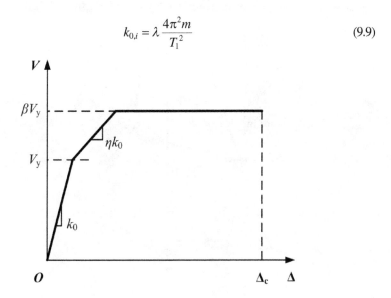

FIGURE 9.3 Backbone curve of the MDOF shear model.

$$V_{y,i} = SA_y \cdot \alpha_1 \cdot m \cdot g \cdot N \cdot \Gamma_i \tag{9.10}$$

$$\eta_I = \frac{SA_u - SA_y}{SD_u - SD_y} \cdot \frac{SD_y}{SA_y} \tag{9.11}$$

$$\beta_i = \frac{SA_u}{SA_y} \tag{9.12}$$

$$\Delta_{c,i} = \delta_{Co} \cdot h \tag{9.13}$$

Where g is the acceleration of gravity; δ_{Co} is the inter-story drift ratio at the threshold of the complete damage state, as suggested by HAZUS (FEMA, 2012a); h is the story height; (SD_y, SA_y) and (SD_u, SA_u) are the yield capacity point and ultimate capacity point, respectively, of the capacity curve suggested by HAZUS (FEMA, 2012a), which are functions of the design intensity and year built; α_1 is the mode factor suggested by HAZUS (FEMA, 2012a); and Γ_i is the ratio between the inter-story shear strength of the ith story, $(V_{y,i})$ and that of the ground story $(V_{y,1})$, which is calculated as follows:

$$\Gamma_i = \frac{V_{y,i}}{V_{y,1}} \tag{9.14}$$

A single-parameter pinching model (Figure 9.4) proposed by Steelman and Hajjar (2009) is adopted to represent the pinching behavior subjected to cyclic loads. Five damage states, ranging from none, slight, moderate, to extensive and complete

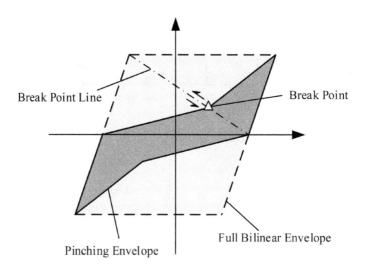

FIGURE 9.4 Diagram of the pinching envelope of the pinching model.

damage, are considered, which are identical to the damage states defined by HAZUS. The inter-story drift ratio is used as the threshold for each structural damage state, and the values for different structural types are based on Table 5.9 of HAZUS (FEMA, 2012a).

The reliability of the city-scale nonlinear time-history analysis method is further validated by comparing the simulation results with earthquake site investigations, experimental results, and a large number of numerical results (Lu & Guan, 2017). With outstanding computational efficiency, this method can be used well for post-earthquake emergency response. There exists an inherent uncertainty in the seismic performance of buildings, which is considered in this method by incorporating the parametric uncertainty of the building backbone curve (Lu et al., 2017b). Consequently, the proposed method can not only provide the building response using the median value of the backbone curve parameters but also provide the responses with the median value ± 1 standard deviation to account for the parametric uncertainty, which is crucial for scientific decision-making.

The nonlinear time-history analysis of the buildings in the target area is implemented using the ground motions obtained from the seismic network. Subsequently, the time histories of the seismic response of each story in every building can be obtained. According to the engineering demand parameters (EDPs) and the damage criteria (Lu & Guan, 2017), the damage state of each building in the region is determined, based on which the destructive power of the ground motion to the target area is evaluated. To make full use of the real-time earthquake ground motions obtained from the densely distributed seismic stations, the destructive powers of ground motions obtained from different seismic stations can be evaluated by inputting the ground motions one-by-one into the building models of the target region. The distribution of building damage ratios under different station records can be given subsequently, which provides an essential reference for post-earthquake rescue work. For example, the destructive power of the ground motions of the August 13, 2018, M5.0 Yunnan Tonghai earthquake can be illustrated intuitively, as shown in Figure 9.5.

The human sense of floor acceleration is highly important in the resilience assessment of communities under moderate seismic actions. Based on the comfort criteria (Simiu & Scanlan, 1996) and floor acceleration computed by the nonlinear time-history analysis, the human sense of different ground motions can be obtained. The distribution of human uncomfortableness under the ground motions of the November 26, 2018, M6.2 Taiwan Strait earthquake is shown in Figure 9.6. Although the damage ratio of buildings under this earthquake is very small, the ratio of human uncomfortableness is still high.

9.2.5 REGIONAL SEISMIC LOSS PREDICTION AND RESILIENCE ASSESSMENT

9.2.5.1 Regional Seismic Loss Prediction Using Conventional Method

Based on the damage state of each building, the seismic loss can be predicted using the conventional loss prediction method. For example, according to the National Standard of China, *"Post-Earthquake Field Works-Part 4: Assessment of Direct*

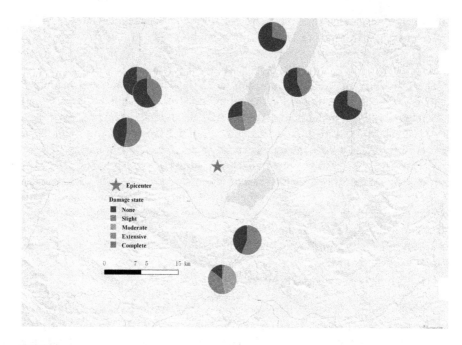

FIGURE 9.5 Destructive power of ground motions of the August 13, 2018, M5.0 Yunnan Tonghai earthquake.

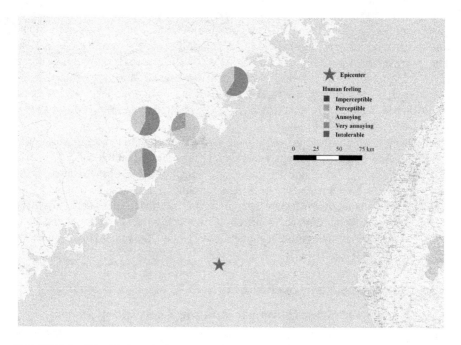

FIGURE 9.6 Distribution of human uncomfortableness under the ground motions of the November 26, 2018, M6.2 Taiwan Strait earthquake.

Loss (GB/T 18208.4-2011)," the house damage loss L_h and decoration damage loss L_d are calculated using Equations 9.15 and 9.16, respectively:

$$L_h = S \times D_h \times P \tag{9.15}$$

$$L_d = \gamma_1 \times \gamma_2 \times (\xi \times S) \times D_d \times (\eta \times P) \tag{9.16}$$

Where S is the building area (m^2); D_h and D_d are the loss ratios of the house and decoration damage, given a damage state; P is the building replacement cost; γ_1 is the correction factor considering different economic conditions of different regions; γ_2 is the building function correction factor; ξ is the proportion of buildings with mid- to-high-quality decoration; and η is the ratio of the building decoration cost to the building construction cost. The values of $\gamma_1, \gamma_2, \xi, \eta$ can be found in Tables A.1–A.4 in GB/T 18208.4 (2011).

9.2.5.2 Regional Resilience Assessment Using FEMA P-58

Furthermore, based on the FEMA P-58 method (next-generation seismic performance assessment method of buildings) and the city-scale nonlinear time-history analysis, a practical approach for regional resilience assessment is proposed (Zeng et al., 2016) to give a more detailed seismic loss and repair time prediction. The main process of the method is as follows: (a) analyze the building response to determine the EDPs on each story of each building using the nonlinear MDOF models, and (b) calculate the economic loss and repair time of components based on the building performance models and the fragility data provided in the FEMA P-58 document. The flowchart of the repair cost calculation is shown in Figure 9.7. Using this method, the damage states of components on different stories can be obtained, and the loss caused by the floor displacement, acceleration, and residual displacement can be considered. Using high-efficiency computational codes, this method can quickly calculate the economic loss and repair time of earthquake-stricken areas.

One of the critical challenges of using the FEMA-P58 method in a region is the assembly of performance models (PGs). The performance model of a building contains its basic information with both structural and nonstructural PGs. The types and quantities of building components can be obtained using the following three

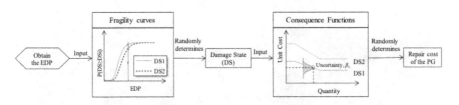

FIGURE 9.7 Calculation of repair cost for a building component using the FEMA-P58 methodology (Zeng et al., 2016).

methods: (a) field survey data and building design drawings, (b) building information models (BIM), and (c) geographic information system (GIS) database.

(a) Field survey data and building design drawings

The type and quantity of each structural PG can be obtained from the structural and architectural drawings of the building. The nonstructural PGs can be determined using the field survey. Note that some effort is required to collect the information. However, the data can be implemented in parallel by groups of people with a basic knowledge of architectural and structural engineering.

(b) Building information models

The detailed building data can be automatically obtained from the BIM in which the building components have different levels of development (LODs). The determination of the component type and the development of a component vulnerability function when the information is incomplete are proposed to produce an acceptable loss prediction (Xu et al., 2019; Zeng et al., 2018). Specifically, the PGs in FEMA P-58 are organized as different classification trees. For example, the classification tree of a gypsum wall board (GWB) partition is shown in Figure 9.8. According to the LODs of the BIMs, all PGs that cannot be determined as leaf nodes in the classification tree due to lack of information will be treated as "potential fragility classifications." A Monte Carlo simulation is subsequently implemented considering all "potential fragility classifications." Consequently, the loss can be predicted even when a very coarse BIM is available. In addition, the modeling rules and information extraction for BIM are proposed to obtain the component information (Xu et al., 2019; Zeng et al., 2018).

(c) GIS database

According to the building inventory of the GIS database, the structural component quantity is estimated based on the statistics from the available literature and design drawings. The nonstructural PG information can be identified according to the normative quantity information provided by Appendix F of FEMA (2012b). Subsequently, the structural and nonstructural PGs can be estimated.

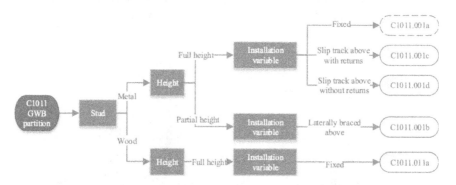

FIGURE 9.8 Classification tree of a gypsum wall board partition component.

A new trend to efficiently and accurately establish the PGs of buildings on a regional scale is using the City Information Model (CIM). CIM is defined as the integration of GIS and BIM (Xu et al., 2014). The CIM of the earthquake-stricken area can be pre-established, which will provide valuable data for the community resilience assessment on different scales.

To demonstrate the resilience assessment method for a region, the seismic economic loss and repair time of Tsinghua Campus (Zeng et al., 2016) were calculated. The ground motion recorded at the Jiuzhaigou Baihe station was input to 619 buildings of Tsinghua Campus. The distribution of the median building loss ratios and repair/rebuilding times are shown in Figure 9.9. The total loss ratio is 0.576%, which is very small. The repair time of the campus is 15 days with parallel repair strategies. The results provide a valuable reference for the resilience assessment of Tsinghua Campus.

9.2.6 High-Performance Computing for Post-Earthquake Emergency Response

High-performance computing (HPC) is incorporated to implement the real-time city-scale THA for the post-earthquake emergency response. The OpenMP library (Chandra et al., 2001) is used to parallelize the code of the city-scale nonlinear time-history analysis. Only 136 s are required to complete the computation of a ground motion record (with an Intel Xeon E5 2630 @2.40 GHz CPU and 64 GB of RAM), which satisfies the requirement of post-earthquake emergency response.

Furthermore, cloud computing is introduced to perform a coarse-grain parallel analysis. Specifically, a number of virtual computers are quickly established (in minutes) on the cloud computing platform (e.g., Aliyun Cloud or Tencent Cloud) after an earthquake. Subsequently, the analyses of different ground motion records are assigned to different virtual computers. Each virtual computer has nearly the same performance as the local computing environment. Thus, regardless of how many ground motions are to be computed, the time consumption is almost the same as that of a single ground motion. Meanwhile, the computational cost is less than one US dollar for each virtual computer. Such a flexible cloud computing platform makes the real-time analysis of a number of ground motions technologically and economically feasible.

9.3 APPLICATIONS IN EARTHQUAKE EMERGENCY RESPONSE

9.3.1 Overview of the Applications

When an earthquake occurs, the ground motion will be collected in a timely manner from the strong ground motion network. The real-time city-scale THA will be conducted for the target region, and the analysis results will be fed back to the decision-makers and reported on the internet in a short time. A program is developed named as "Real-time Earthquake Damage Assessment using City-scale Time-history analysis" ("RED-ACT" for short) to automatically implement the above workflow. To date, the RED-ACT system has been applied to several earthquakes in China and other countries around the world, as listed in Table 9.1.

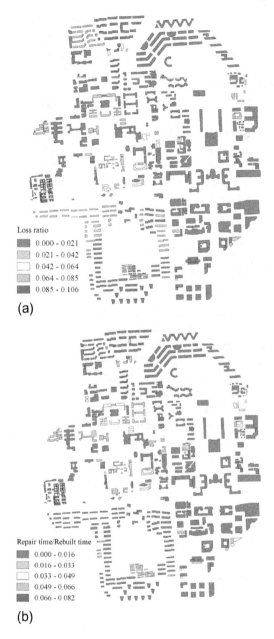

FIGURE 9.9 Distribution of median building loss ratios and repair/rebuild time for Tsinghua Campus.

TABLE 9.1
Applications of the RED-ACT System

ID	Earthquake name	ID	Earthquake name
1	2016-12-08 M6.2 Xinjiang Hutubi earthquake	18	2018-11-26 M6.2 Taiwan Strait earthquake
2	2016-12-18 M4.3 Shanxi Qingxu earthquake	19	2018-12-08 M4.5 Xinjiang Changji earthquake
3	2017-03-27 M5.1 Yunnan Yangbi earthquake	20	2018-12-16 M5.7 Sichuan Yibin earthquake
4	2017-08-08 M7.0 Sichuan Jiuzhaigou earthquake	21	2018-12-20 M5.2 Xinjiang Kizilsu earthquake
5	2017-09-30 M5.4 Sichuan Qingchuan earthquake	22	2019-01-03 M5.3 Sichuan Yibin earthquake
6	2018-02-06 M6.5 Taiwan Hualien earthquake	23	2019-01-07 M4.8 Xinjiang Jiashi earthquake
7	2018-02-12 M4.3 Hebei Yongqing earthquake	24	2016-04-16 M7.3 Japan Kumamoto earthquake
8	2018-05-28 M5.7 Jilin Songyuan earthquake	25	2016-08-24 M6.2 Italy earthquake
9	2018-08-13 M5.0 Yunnan Tonghai earthquake	26	2016-11-13 M8.0 New Zealand earthquake
10	2018-08-14 M5.0 Yunnan Tonghai earthquake	27	2017-09-20 M7.1 Mexico earthquake
11	2018-09-04 M5.5 Xinjiang Jiashi earthquake	28	2017-11-23 M7.8 Iraq earthquake
12	2018-09-08 M5.9 Yunnan Mojiang earthquake	29	2018-06-18 M6.1 Japan Osaka earthquake
13	2018-09-12 M5.3 Shanxi Ningqiang earthquake	30	2018-09-06 M6.9 Japan Hokkaido earthquake
14	2018-10-16 M5.4 Xinjiang Jinghe earthquake	31	2018-10-26 M5.4 Japan Hokkaido earthquake
15	2018-10-31 M5.1 Sichuan Xichang earthquake	32	2018-12-01 M7.0 Alaska earthquake
16	2018-11-04 M5.1 Xinjiang Atushi earthquake	33	2019-01-03 M6.2 Japan Kumamoto earthquake
17	2018-11-25 M5.1 Xinjiang Bole earthquake		

9.3.2 2017 M7.0 JIUZHAIGOU EARTHQUAKE

The seismic damage assessment of the 2017 Jiuzhaigou earthquake is a typical application case (Lu et al., 2017a). After the earthquake, several sets of ground motion records were obtained from the seismic network, and the seismic damage prediction of the target region was completed in 2 hours (including the time for the data checking and report editing/publishing) by using the method proposed. The predicted damage of a typical town and country in the Aba region under the ground motion of the Jiuzhaigou Baihe station is shown in Figure 9.10. The predicted

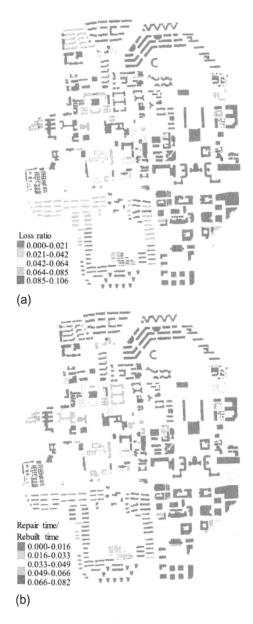

FIGURE 9.10 Seismic results of a typical town and country in the Aba region subjected the ground motion from the Jiuzhaigou Baihe station.

results show that the buildings in the disaster area may be damaged to some extent, but the ratio of collapse is very small, which is consistent with the actual post-earthquake site investigations (Dai et al., 2018). The results provide a useful reference for the earthquake emergency response and scientific decision-making of earthquake disaster relief.

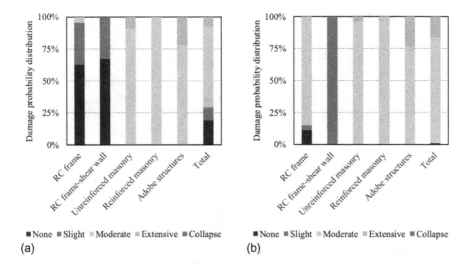

FIGURE 9.11 (a) Location of the 8047 ground motion station (CESMD, 2019), and the ground motion recorded by the 8047 station: (b) EW, (c) NS, (d) UD.

9.3.3 2018 Mw 7.0 Anchorage Earthquake

The seismic damage assessment of the 2018 Mw 7.0 Anchorage earthquake is another typical application case (Lu et al., 2018). On November 30, 2018 (local time), an Mw 7.0 earthquake occurred in Alaska, the United States. The epicenter was at 61.35 N, 150.06 W with a depth of 40 kilometers (StEER & EERI, 2018). Six ground motions of the Alaska earthquake event were recorded. The ground motions recorded at the 8047 station (61.189 N, 149.802 W, shown in Figure 9.11) is a typical ground motion. The peak ground accelerations (PGAs) of horizontal and vertical components of the 8047 ground motion were 807.162 cm/s^2 and 367.243 cm/s^2, respectively. The ground motions are shown in Figure 9.11.

Using the ground motions obtained from the strong motion networks and the city-scale nonlinear time-history analysis, the "RED-ACT" system predicted the damage ratio and human uncomfortableness distribution of the buildings near different stations in less than 1 hour, as shown in Figures 9.12 and 9.13. The post-earthquake investigation showed that this Mw 7.0 earthquake produced less-than-expected damage to buildings, including businesses, homes, and schools in downtown Anchorage, with most damage limited to nonstructural elements and contents (StEER & EERI, 2018), which is consistent with the prediction given by the proposed method.

9.4 CONCLUSIONS

Based on the city-scale nonlinear time-history analysis and the regional seismic loss prediction, a real-time city-scale time-history analysis method is proposed in this work. A program named "RED-ACT" was developed to automatically implement the described workflow. The reliability and advantages of the proposed method in

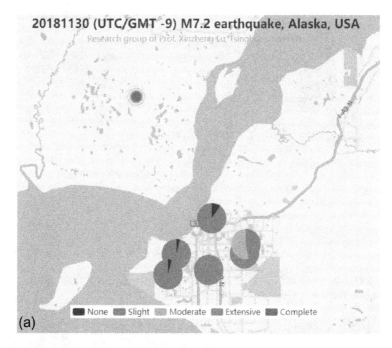

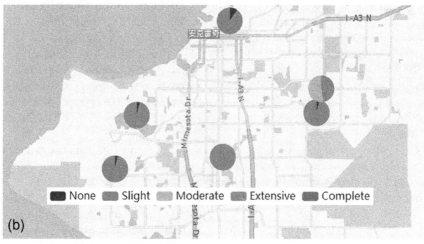

FIGURE 9.12 Damage ratio distribution of the buildings near different stations of the 2018 Mw 7.0 Anchorage earthquake.

this work were demonstrated through actual earthquake events. Then the program was applied to various earthquake events. The main conclusions are as follows:

(1) The uncertainty problem of ground motion input is solved properly with the proposed method based on the real-time ground motion obtained from the seismic stations.

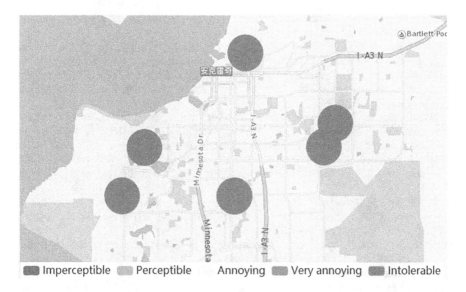

FIGURE 9.13 Distribution of human uncomfortableness under the ground motions of the 2018 Mw 7.0 Anchorage earthquake.

(2) The amplitude, spectrum, and duration characteristics of ground motions, as well as the stiffness, strength, and deformation characteristics of different buildings are fully considered in this method based on the nonlinear time-history analysis and MDOF models.

(3) Using the real-time city-scale time-history analysis and the corresponding report system, the assessment of the earthquake's destructive power, repair time, and economic loss can be obtained shortly after an earthquake event, which provides a useful reference for scientific decision-making for earthquake disaster relief. This work is highly significant for enhancing the resilience of earthquake-stricken areas.

ACKNOWLEDGMENTS

The authors are grateful for the financial support received from the National Key R&D Program (number 2018YFC1504401) and the ground motion data provided by the China Earthquake Networks Center, the Center for Engineering Strong Motion Data (CESMD), and K-NET and KiK-net.

REFERENCES

American Society of Civil Engineers (ASCE), (2010), *Minimum design loads for buildings and other structures (ASCE/SEI 7–10)*, Reston, VA: American Society of Civil Engineers (ASCE).

Center for Engineering Strong Motion Data (CESMD), (2019), [Online]. Available: https://www.strongmotioncenter.org/cgi-bin/CESMD/iqrStationMap.pl?ID=us1000hyfh.

Chandra, R., Dagum, L., Kohr, D., Menon, R., Maydan, D., McDonald, J., (2001), *Parallel programming in OpenMP*, Morgan Kaufmann.

Dai, J. W., Sun, B. T., Li, S. Y., Tao, Z. R., Ma, Q., Zhang, L. X., Lin, J. Q., (2018), *Engineering damage in Jiuzhaigou M 7.0 earthquake*, Seismological Press, 19. (in Chinese).

Erdik, M., Şeşetyan, K., Demircioğlu, M. B., Hancılar, U., Zülfikar, C., (2011), "Rapid earthquake loss assessment after damaging earthquakes," *Soil Dynamics and Earthquake Engineering*, 31(2), 247–266.

Federal Emergency Management Agency (FEMA), (2012a), *Multi-hazard loss estimation methodology-earthquake model technical manual (HAZUS-MH 2.1)*, Final report, Washington, DC: Federal Emergency Management Agency.

Federal Emergency Management Agency (FEMA), (2012b), "Seismic performance assessment of buildings volume 1-methodology", Technical report FEMA-P58, Washington, DC.

GB50009-2012, (2012), *Load code for design of building structures (GB 50009–2012)*, Beijing, China: Ministry of Housing and Urban-Rural Development of the People's Republic of China (MOHURD).

GB50010-2010, (2010), *Code for seismic design of concrete structures (GB50010-2010)*, Beijing, China: Ministry of Housing and Urban-Rural Development of the People's Republic of China (MOHURD).

GB50011-2010, (2010), *Code for seismic design of buildings (GB50011-2010)*, Beijing, China: Ministry of Housing and Urban-Rural Development of the People's Republic of China (MOHURD).

GB/T18208.4-2011, (2011), "Post-earthquake field works-part 4: assessment of direct loss, GB/T 18208.4-2011," Standardization Administration of China, Beijing, China. (in Chinese).

Lu, X. Z., (2018), "Damage capacity of the ground motions of the Dec. 1, Alaska Earthquake," Tsinghua University, Research Group Report.

Lu, X. Z., Guan, H., (2017), *Earthquake disaster simulation of civil infrastructures: from tall buildings to urban areas*, Springer.

Lu, X. Z., Han, B., Hori, M., Xiong, C., Xu, Z., (2014), "A coarse-grained parallel approach for seismic damage simulations of urban areas based on refined models and GPU/CPU cooperative computing," *Advances in Engineering Software*, 70, 90–103.

Lu, X., Cheng, Q., Xu, Z., and Sun, C., (2019), "Real-time city-scale time-history analysis and its application in resilience-oriented earthquake emergency responses," *Applied Sciences*, 9(17), 16.

Lu, X. Z., Tian, Y., Guan, H., Xiong, C., (2017b), "Parametric sensitivity study on regional seismic damage prediction of reinforced masonry buildings based on time-history analysis," *Bulletin of Earthquake Engineering*, 15(11), 4791–4820.

Population Census Office under the State Council, Department of Population and Employment Statistics of National Bureau of Statistics, (2012), *Tabulation on the 2010 population census of the People's Republic of China*, China Statistics Press. (in Chinese).

Simiu, E., Scanlan, R. H., (1996), *Wind effects on structures: fundamentals and applications to design*, 3rd edition, New York: John Wiley.

Sobhaninejad, G, Hori, M, Kabeyasawa, T., (2011), "Enhancing integrated earthquake simulation with high performance computing," *Advances in Engineering Software*, 42(5), 286–292.

Steelman, J. S, Hajjar, J. F., (2009), "Influence of inelastic seismic response modeling on regional loss estimation," *Engineering Structure*, 31(12), 2976–2987.

Structural Extreme Event Reconnaissance Network (StEER), Earthquake Engineering Research Institute (EERI), (2018), "Alaska earthquake preliminary virtual assessment team (P-VAT) joint report," December 6.

Xiong, C., Lu, X. Z., Guan, H., Xu, Z., (2016), "A nonlinear computational model for regional seismic simulation of tall buildings," *Bulletin of Earthquake Engineering*, 14(4), 1047–1069.

Xiong, C., Lu, X. Z., Lin, X. C., Xu, Z., Ye, L. P., (2017), "Parameter determination and damage assessment for THA-based regional seismic damage prediction of multi-story buildings," *Journal of Earthquake Engineering*, 21(3), 461–485.

Xu, X., Ding, L. Y., Luo, H. B., Ma, L., (2014), "From building information modeling to city information modelling," *Journal of Information Technology in Construction*, 19, 292–307.

Xu, Z., Lu, X., Zeng, X., Xu, Y., Li, Y., (2019), "Seismic loss assessment for buildings with various-LOD BIM data," *Advanced Engineering Informatics*, 39, 112–126.

Zeng, X, (2018), "Direct seismic loss assessment and fire following earthquake simulation of urban buildings," Doctoral thesis, Tsinghua University, Beijing. (in Chinese).

Zeng, X., Lu, X. Z., Yang, T. Y., Xu, Z., (2016), "Application of the FEMA-P58 methodology for regional earthquake loss prediction," *Natural Hazards*, 83(1), 177–192.

10 Functionality Analyses of Engineering Systems
One Step toward Seismic Resilience

Tao Wang, Qingxue Shang, and Jichao Li

CONTENTS

10.1 Introduction 163
10.2 A Quantitative Framework for Seismic Resilience Assessment.................... 164
10.3 System Model and System Assessment 166
10.4 Case Study of Emergency Functionality in a Hospital................................. 167
10.5 Discussion and Conclusions.. 174
10.6 Acknowledgements.. 175
References... 175

10.1 INTRODUCTION

The 2005 World Conference on Disaster Reduction (WCDR) confirmed the importance of the term resilience on disaster mitigation. Since then, the concept of resilience has been gaining extensive attention in the earthquake engineering community, including the policy makers, disaster managers, engineers, and scientists. The Federal Emergency Management Agency (FEMA) initiated a series of mitigation programs designed to reduce future earthquake losses by establishing Project Impact in 1997 (FEMA, 2000). FEMA also instituted a Disaster Resistant Universities program to improve the ability to withstand the effects of probable hazard events without unacceptable losses or interruptions (FEMA, 2003), specifically in terms of repair costs and downtime. The National Research Council (NRC) of the United States established the Committee on Earthquake Resilience-Research, Implementation, and Outreach in 2011. In the report of the NRC, seismic resilience is intended to be a future direction in earthquake engineering (NRC, 2011). In 2013, San Francisco became one of the first 100 Resilient Cities to receive funding and support from the Rockefeller Foundation to develop and implement policies and programs to improve San Francisco's overall resilience. After the 2011 earthquake off the Pacific coast of Tōhoku, resiliency has been deemed more urgent and the Business Continuity Plan (BCP) has become one of the main research objectives in Japan.

Seismic resilience is defined as the ability of an engineering system to resist, restore, and adapt to an earthquake impact (Bruneau et al., 2003), as shown in

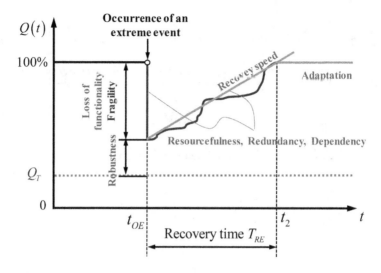

FIGURE 10.1 Typical model of seismic resilience.

Figure 10.1. The recovery capacity is quantified as the variation of the functionality over time. Most of the evaluation frameworks or methods have been conceptual and semi-quantitative up to now. A quantitative framework for seismic resilience assessment and a system model are necessary to help quantify resilience. In this chapter, a quantitative framework accompanied by a system model is proposed to evaluate resilience.

10.2 A QUANTITATIVE FRAMEWORK FOR SEISMIC RESILIENCE ASSESSMENT

The proposed framework to quantify the seismic resilience of an engineering system is illustrated in Figure 10.2. The framework consists of four steps, i.e., the seismic hazard analysis, the fragility analysis, the seismic risk analysis, and the calculation of the seismic resilience based on system analysis. The first three steps are very similar to those used to evaluate the seismic risk of an engineering system. The final step is to assess the resilience using the seismic resilience evaluation analysis (SREA). Various indices have been proposed in past studies to define the functionality, such as economic loss and downtime. The functionality of an engineering system actually has multiple meanings. It is difficult to quantify it using a single index. In this study, the economic loss caused by the related components is used to quantify the corresponding functionality. Therefore, it is very important to quantify the importance factor of each component which determines the contribution of each component to the functionality loss of a system. In doing so, the seismic resilience can be explicitly quantified by the loss and recovery time. The casualty index is also important and can be incorporated into the building resilience model. However, more effort is needed to accurately define the casualty model, and this is beyond the scope of this chapter.

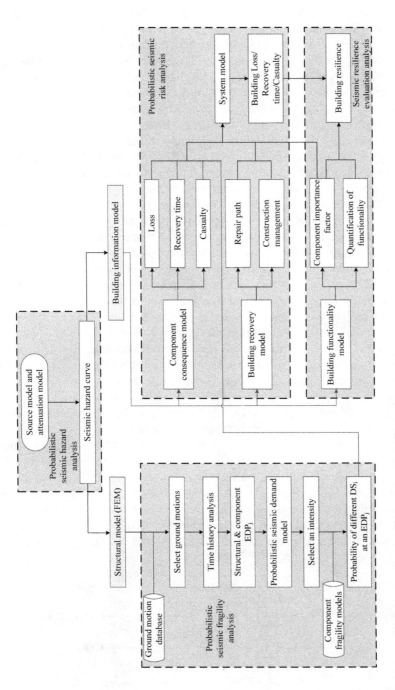

FIGURE 10.2 Quantitative framework for seismic resilience assessment.

10.3 SYSTEM MODEL AND SYSTEM ASSESSMENT

In the process of SREA, a system model and a new method called the state tree method will be adopted. Complex systems are often organized, structured, and redundant. A multiple-input and multiple-output fundamental system (Figure 10.3) is used to illustrate the state tree methods. The system is viewed as a "plant" that consumes "resources" and provides "services." The blocks with numbers represent the system's components, which can be further categorized into sub-systems, some of which deal with input resources and some with output services. Therefore, the functionality of this fundamental system can be defined by the services and quantified as either output 1 or output 2, or both output 1 and output 2. The successful services of a system also depend upon the available resources. The availability of resources can also be quantified in a manner similar to that of the services, that is, input 1, input 2, or both input 1 and input 2.

One of the most challenging tasks in system assessment is to evaluate the operational state of a given system. In the proposed method, the system state is evaluated directly and simply through the state tree once the components' states are determined. The procedure progresses as follows. The operational state of one component is recognized as 0 (failure) or 1 (success). The operational states of fictional events (FEs) can be evaluated according to the row vectors that represent the operational state combinations of the involved components or FEs. For example (Figure 10.4), suppose the operational state of components 1, 3, 4, and 5 are 1, 1, 0, and 1, respectively. From the row vector of [0 1], representing the state combination of components 4 and 5, the state of "FE 4" can be evaluated as success. The operational state of "FE 3" can be continuously evaluated as a success because the two events under the "OR" gate, that is, component 3 and "FE 4" are both successful. The operational state of FE 1 is thus successful because another row vector of [1 1] representing both component 1 and "FE 3" is successful. Similarly, the state of "FE 2" can also be determined, and the operational state of the entire concerned system can be evaluated.

The system's failure probability is calculated with the Monte-Carlo simulation, which is an effective numerical method to solve problems with high nonlinearity that are difficult to solve by any analytical method. The procedure of a typical Monte-Carlo simulation is shown in Figure 10.5. In one simulation, a random number uniformly distributed between 0 and 1 is generated for each component at the bottom level of the system model. If the generated number is less than or equal to the failure probability defined by the corresponding fragility curve at a given EDP, the

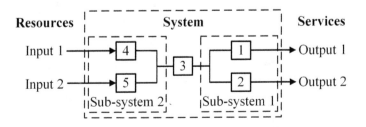

FIGURE 10.3 Fundamental multiple-input and multiple-output system.

Functionality Analyses

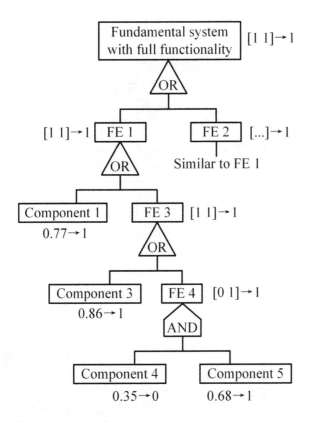

FIGURE 10.4 State evaluation using state tree.

component is considered to fail, otherwise it succeeds. For the example shown in Figure 10.4, assume the failure probability of each component is 0.5 at a PGA of 0.5 g, the operational state of components 1, 3, 4, and 5 can be evaluated directly according to the corresponding random numbers 0.77, 0.86, 0.35, and 0.68, respectively. Once the states of all components are determined, the system's operational state can be calculated using the procedure described in Section 2.5. For a certain value of EDP, the simulation will be repeated n times using different random numbers, and the system's failure probability can be calculated as the ratio of failed numbers to the number of simulations. Such simulations will then be repeated m times for different values of EDP, and the fragility of the system can be obtained.

10.4 CASE STUDY OF EMERGENCY FUNCTIONALITY IN A HOSPITAL

The proposed methodological framework is demonstrated with emergency functionality analysis and resilience assessment on a hospital building. For emergency functionality, the state tree model can be developed based on the fault tree model mentioned above. Several fictional events are added to represent the states of component groups (Figure 10.6).

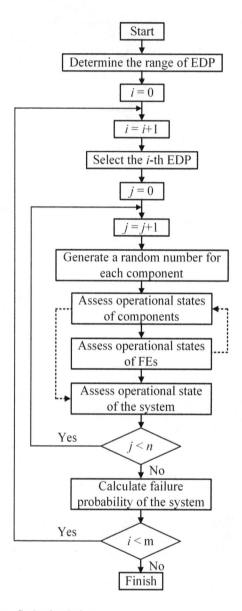

FIGURE 10.5 Monte-Carlo simulation.

Hospital systems are designed to provide medical care for patients. Emergency functionality of an emergency department is defined as the ability to provide emergency medical care and is calculated as a normalized number of working lines (percentage of full ED operation), as presented in Equation (10.1), where N_{total} is the number of possible working lines which is the reference functionality before the earthquake or under normal operational state and N_{total} is specified as four in this study, $N(t)$ is the number of working lines at a given PGA and is calculated based on

Functionality Analyses

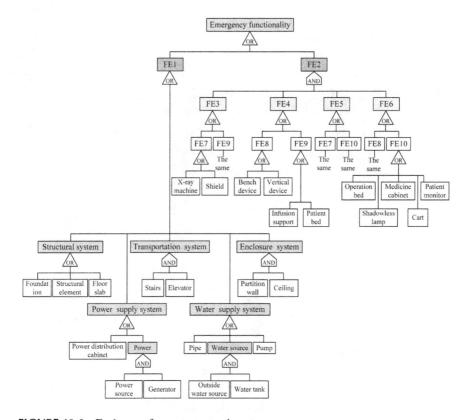

FIGURE 10.6 Fault tree of an emergency department.

full probability theory, as given in Equation (10.2), $Q(t)$ represents the functionality of a system at time t. $P[N = i \mid PGA = pga]$ is the probability of an event when the number of working lines available is i at a given PGA and is calculated as the difference between the probabilities of exceeding a maximum of i and $i + 1$ working lines.

$$Q(t) = \frac{N(t)}{N_{total}} \tag{10.1}$$

$$N(t) = \sum_{i=1}^{N_{total}} i \cdot P[N = i \mid PGA = pga] \tag{10.2}$$

Seismic resilience is a dynamic process of the functionality variation over time for an engineering system after earthquakes. Functionality reduction will happen when the earthquake hits and the functionality will increase when the damaged components of the system have been repaired. The repair path of a system can be a complex issue in considering the construction management strategy. For simplicity, in this study, an idealized repair path considering the component location and construction sequence is used. As depicted in Figure 10.7, the idealized repair

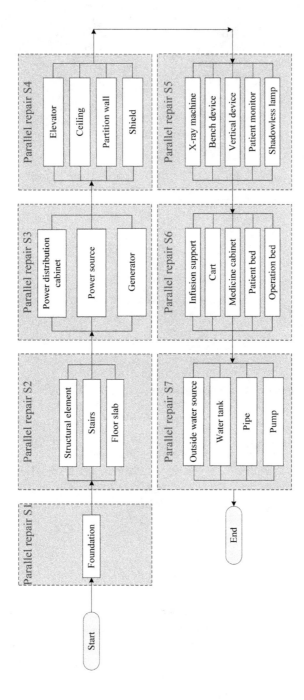

FIGURE 10.7 Idealized repair path for emergency functionality.

Functionality Analyses

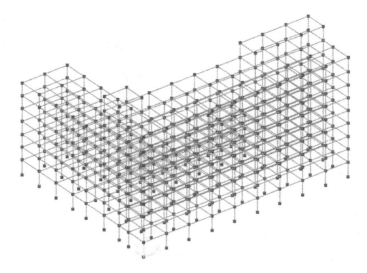

FIGURE 10.8 Three-dimensional model of the hospital building.

path has seven steps that occur in series, whereas in each step, several component repairs can proceed in parallel. For one parallel repair sequence, the recovery time is determined by the component which takes the largest time to restore. Once the recovery time of each step is obtained, the recovery time of the system is calculated as the sum of the recovery times of all seven steps. It is assumed that the recovery work would be conducted in all stories simultaneously and the recovery time of one specific component is the maximum time needed for repairing that component in each story.

The hospital building used for the case study was a ten-story reinforced concrete (RC) frame (Figure 10.8) designed following the Chinese Code for Seismic Design of Buildings (GB 50011-2010, 2010) as presented by Wen et al. (2019). The results of the probabilistic seismic demand model (PSDM) for maximum inter-story drift ratio and peak floor acceleration (Wen et al., 2019) were used in this chapter to evaluate the fragility of structural components and nonstructural components in the case study hospital.

In the Monte-Carlo simulation, the PGA value ranged from 0.01 g to 1.0 g with the interval of 0.01 g, and the simulation was repeated 1000 times for each PGA value. The fragility curves for $F = 0\%, 25\%, 50\%, 75\%$ are shown in Figure 10.9. Fragility parameters fitted by MATLAB (The MathWorks Inc., 2016) are presented in Table 10.1. The median capacities with a 50% failure probability for $F = 0\%, 25\%, 50\%, 75\%$ are all less than the median capacity of the single components, which indicates that the seismic reliability of the emergency functionality is lower than the individual components and the interdependency between different components and their influence to the functionality of a system must be involved in seismic performance assessment.

Residual functionality $Q(t = t_0)$ is the functionality after an earthquake when the recovery process has not been conducted. Residual functionality was calculated

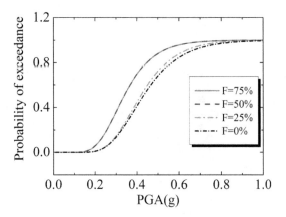

FIGURE 10.9 System fragility curves.

TABLE 10.1
Fragility Parameters (Unit: g)

Fragility	Median	Beta
$F = 0\%$	0.3383	0.3244
$F = 25\%$	0.3383	0.3244
$F = 50\%$	0.4152	0.3154
$F = 75\%$	0.4282	0.3348

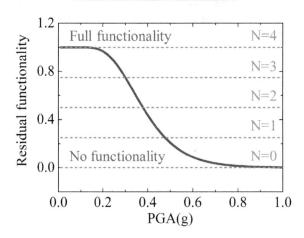

FIGURE 10.10 Residual functionality for different PGA values.

using Equation (10.1) and was given in Figure 10.10. Different levels of functionality, namely, no functionality, low functionality, slight functionality, moderate functionality, and full functionality were also presented in Figure 10.10. The level of functionality can be used to reflect the redundancy of the system under different earthquake intensities.

Functionality Analyses

The functionality of the system after each recovery process was presented in Figure 10.11, where Si represents the repair sequences $S1, S2, \ldots,$ and Si ($i = 1,2,\ldots,6,7$) have been finished. The functionality will increase after each repair sequence and comes to 1.0 (system performance level before the earthquake happened) when the seven recovery sequences have been finished.

Using the linear recovery model suggested by Cimellaro et al. (2010a), the resilience curves for the different earthquake intensities were developed considering the repair path and are presented in Figure 10.12. Resilience index R_{Resi} can be determined by Equation (10.3) as suggested by Cimellaro et al. (2010b), where $Q(t)$ is system functionality, t_{OE} is the time of occurrence of the earthquake, T_{RE} is the recovery time. The resilience indexes of the different earthquake intensities are listed in the last row of Table 10.2. Once an earthquake occurs, the functionality of the system is affected and decreases by 0.000048%, 3.28%, and 56.24% for the SLE, DBE, and MCE, respectively. The recovery time of the emergency functionality meets the 8.13-day resilience demand (Wen et al., 2019) for the SLE and DBE intensities.

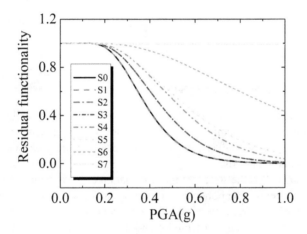

FIGURE 10.11 Functionality after each recovery process.

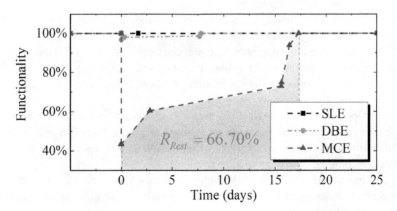

FIGURE 10.12 Resilience curves under different earthquake intensities.

TABLE 10.2
Resilience of the System Under Different Earthquake Intensities

Earthquake intensity	SLE	DBE	MCE
PGA (g)	0.07	0.2	0.4
Sa (g)	0.0513	0.144	0.2883
Recovery time (days)	1.69	7.81	17.29
Functionality loss	0.000048%	3.28%	56.24%
R_{Resi}	99.99995%	98.33%	66.70%

But the resilience of the emergency functionality still needs to be improved to meet the demand under MCE.

$$R_{\text{Resi}} = \frac{1}{T_{RE}} \int_{t_{OE}}^{t_{OE}+T_{RE}} Q(t)dt \qquad (10.3)$$

10.5 DISCUSSION AND CONCLUSIONS

This chapter proposes a framework for quantifying the seismic resilience of engineering systems, which is composed of probabilistic seismic hazard analysis, probabilistic seismic fragility analysis, probabilistic seismic state analysis, and seismic resilience evaluation analysis.

A new method called the state tree method that takes into account the interdependency between different components and their influence on the functionality is used to measure the seismic fragility of an emergency department. The state tree model is developed based on a fault tree model and a success path method. The interactions and coupling effect of shared components in analyzing system fragility are uncoupled by the state tree method.

System fragility is defined as consistent with the definition of component fragility based on four working lines identified by the state tree model. Emergency functionality is defined as the ability to provide emergency medical care and is quantified by the normalized number of working lines based on full probability theory. An idealized repair path considering the component location and construction sequence is used to calculate the system recovery time.

The proposed methodological framework is demonstrated with emergency functionality analysis and resilience assessment of a case study hospital building. Functionality loss and recovery time, which resulted from each repair sequence, were obtained based on the simulation results.

In conclusion, the proposed methodology can be used to conduct fragility analysis and resilience assessment for emergency functionality of hospital systems and any other engineering system.

10.6 ACKNOWLEDGEMENTS

This research was funded by the Scientific Research Fund of Institute of Engineering Mechanics, CEA (2016A06). Any opinions, findings, and conclusions or recommendations expressed in this chapter are those of the authors and do not necessarily reflect the views of the sponsors.

REFERENCES

Bruneau M., Chang S. E., Eguchi R. T., et al. (2003), "A framework to quantitatively assess and enhance the seismic resilience of communities," *Earthquake Spectra*, 19(4): 733–752.

Cimellaro G. P., Reinhorn A. M., Bruneau M. (2010a), "Seismic resilience of a hospital system," *Structure & Infrastructure Engineering*, 6(1–2): 127–144.

Cimellaro G. P., Reinhorn A. M., Bruneau M. (2010b), "Framework for analytical quantification of disaster resilience," *Engineering Structures*, 32(11): 3639–3649.

Federal Emergency Management Agency (2000), *Planning for a Sustainable Future: The Link Between Hazard Mitigation and Livability*. Washington, DC: Federal Emergency Management Agency.

Federal Emergency Management Agency (FEMA) (2003), *Building a Disaster-Resistant University*. Washington, DC: Federal Emergency Management Agency.

GB 50011-2010 (2010), *Code for Seismic Design of Buildings*. Beijing: China Architecture and Building Press (in Chinese).

Li J. C., Wang T., Shang Q. X. (2019), "Probability-based seismic reliability assessment method for substation systems," *Earthquake Engineering and Structural Dynamics*, 48: 328–346.

National Research Council (NRC) (2011), *National Earthquake Resilience: Research, Implementation, and Outreach*. Washington, DC: National Academies Press.

Tavares D. H., Padgett J. E., Paultre P. (2012), "Fragility curves of typical as-built highway bridges in eastern Canada," *Engineering Structures*, 40(7): 107–118.

The MathWorks, Inc. (2016), Matlab, Version R2016b. Natick, MA: The Math Works.

Wen, W., Ji, D., Zhai, C., and Xie, L., (2019), "A framework to assess the seismic resilience of urban hospitals," *Advances in Civil Engineering. doi.org/10.1155/2019/7654683*.

11 Resilience of Bridges in Infrastructural Networks

Marco Domaneschi

CONTENTS

11.1 Introduction 177
11.2 The Applied Element Method 179
11.3 The Bridge Structure 180
11.4 The Bridge Model 180
11.5 Robustness and Redundancy Issues 181
11.6 Seismic Response Simulation of The Yong-He Bridge: Original and Modified Configuration 182
11.7 General Remarks 182
11.8 Benefits from Structural Control 183
11.9 Benefits from Structural Health Monitoring 184
References 186

11.1 INTRODUCTION

The recent disasters of the disproportionate collapse of the Morandi Bridge in Genoa (August 14, 2018) and a bridge in Kolkata (India, September 4, 2018) have highlighted the importance of building resilient structures for our communities. Concepts such as redundancy and robustness play a significant role, respectively, to provide alternative resources in the event of local out-of-service of structural elements, and to sustain performance or stress levels without showing degradation or a loss of functionality.

Benefits can also be obtained from the implementation of intelligent control systems to compensate for any out-of-service structural elements through the parameter modifications of intelligent components. Furthermore, monitoring systems may also be useful to obtain information on the degradation conditions, to be able to adopt the necessary actions in advance and, thus, reduce the risks of disproportionate collapse.

Robustness of buildings and bridges is defined as the ability of the structure to withstand a given level of stress or demand (e.g., damage) without suffering degradation or loss of function. Besides, *redundancy* is another structural characteristic that is often required at the design level for the benefits it provides against unwanted behaviors. Redundancy is defined as the quality of having alternative paths in the structure by which the forces can be transferred, which allows the structure to remain stable following the failure of any single element (Cimellaro et al. 2010).

Such characteristics, whose interconnection has also been recognized by Kanno and Ben-Haim (2011), are desirable in structural systems that are able to reduce vulnerability and therefore avoid the disproportionate collapse. Disproportionate collapse occurs when an initial local failure that is produced by a small triggering event leads to the widespread failure of other structural components such that the structure collapses. It is also referred to as a progressive collapse (Domaneschi et al. 2018, 2019).

In recent years several studies have been developed on structural collapses, but the most attention has been given to buildings, leaving the bridges still uncharted or partially investigated by only a few researchers. However, recent events of bridge collapses, namely in Genoa (Italy) on August 14, 2018, and in Kolkata (India) on September 4, 2018, have focused the public interest on infrastructure safety for its consequences in terms of fatalities and injuries, but also of the economy and social losses (Domaneschi et al. 2018, 2019).

The General Services Administration guidelines and the Unified Facilities Criteria are the two most important guidelines that address the progressive collapse of structures. However, they are focused on buildings and the progressive collapse of bridges is only briefly outlined in the guidelines, e.g., according to the Posttensioning Institute, the sudden loss of any one cable must not lead to the rupture of the entire structure (Domaneschi et al. 2018, 2019).

Among the recent investigations, the structural behavior of a long-span suspension bridge segmented by zipper-stoppers after the sudden rupture of some of its cables was studied by Shoghijavan and Starossek (2018a). It was found that increasing the robustness of the structural system through segmentation was a possible approach to prevent the progressive collapse of bridges due to cable failure.

The structural behavior of long-span suspension bridges after the sudden rupture of a cable was studied by Shoghijavan and Starossek (2018b, 2018c). The load carried by the failed cable must be redistributed to the other structural components and the cables adjacent to the failed cable receive most of the redistributed load and become the critical member. Furthermore, cable failure produces large bending moments on the girder of the bridge. With the aim of studying these behaviors, a comprehensive analytical approach is proposed.

Moving to the cable-stayed-bridge class, Wang et al. (2017) investigate the collapse of a cable-stayed bridge due to strong seismic excitations, simulating the structural response through the explicit dynamic finite element method. They identified that the main reason for the collapse of cable-stayed bridges was in the failure of piers and pylons, rather than the failure of cables or main girder components. The cable-stayed bridge collapse mechanism under strong earthquake excitations was also investigated by Zong et al. (2016). The results indicated that the ground motion with the longest predominant period caused the collapse of the bridge. The introduction of viscous dampers at the connections of the pylons and main girder can enhance the earthquake-resistant collapse capacity of the bridge.

Das et al. (2016) introduced the alternate path method to cable-stayed bridges to help prevent their progressive collapse. The structural response is discussed for multiple types of cable-loss to recognize the lack of robustness in the structure and to suggest more robust design options.

Wolff and Starossek (2009, 2010) investigated the disproportionate collapse behavior of a cable-stayed bridge within a cable-loss scenario. The failure of three adjacent cables that stabilized the bridge girder in compression were responsible for the deck buckling as a result of the high normal forces.

The importance of providing system redundancy was also highlighted by the collapse of the Mississippi River Bridge in Minneapolis (Salem and Helmy 2014), Minnesota, in 2007, in which the whole bridge, which had been classified as non-redundant by the National Transportation Safety Board, catastrophically collapsed after the failure of a gusset plate connection.

The response of a current cable-stayed bridge with respect to the issue of disproportionate collapse through nonlinear dynamic analysis and the use of the applied element method (AEM) was analyzed by Domaneschi et al. (2018, 2019). Furthermore, new redundancy indexes that account for the system's reserve resources and quantitatively allow for the evaluation of alternative structural configurations were proposed. In detail, a model of an existing bridge that has been widely examined in the literature through an international benchmark study on structural monitoring and control is implemented through the AEM and then validated. Solutions and interventions in order to avoid disproportionate collapse and to increase bridge robustness and redundancy is analyzed and discussed. It consists of a new cables-configuration and deck strengthening.

Besides, positive contributions to the risk reduction in bridge structures of transportation networks may come from structural control and health monitoring (SHM) solutions. With a focus on intelligent control systems, these procedures pay special attention to rapidity, looking at the time interval required to mitigate losses incurred from bridge damage during a seismic event, since it is an essential aspect of resilient structural behavior. Robustness and redundancy are also key dimensions of resilience that can be improved by the adoption of such techniques. Therefore, an automatic intervention through a semi-active system can be developed on a bridge structure to demonstrate the positive outcomes of intelligent control solutions.

When SHM solutions are considered, they are connected to the innovative view of resilience for civil structures and infrastructures, which, in recent years, has attracted the attention of several specialists, also in the bridge engineering field. The time required to recover the original performance of the undamaged structure, which is identified as the rapidity dimension of resilience, is becoming a new interesting and important aspect of investigation into existing and new constructions to reduce the cost of failures. This is obviously based, first of all, on the capacity to assess if damage has occurred. SHM can give fundamental benefits in this respect. Some recent results are also briefly presented with reference to some relevant researches.

11.2 THE APPLIED ELEMENT METHOD

The presented study investigates the progressive collapse of a real bridge through nonlinear dynamic analysis and an AEM based software (Applied Science International 2017). The AEM is an innovative modeling method that adopts the concept of discrete cracking. Through two decades of continuous development, AEM was proven to be the method that can track the structural collapse behavior passing through all stages of the application of loads: elastic stage, crack initiation and propagation in tension-weak materials, reinforcement yielding, element separation, element collision (contact), and collision with the ground and with adjacent structures.

Within the AEM, the structure is modeled as an assembly of small elements that virtually subdivide the structure. A couple of elements are assumed to be connected by one normal and two orthogonal shear springs distributed on the elements adjacent faces (Domaneschi et al. 2018, 2019). Each group of springs completely represents stresses and deformations for a constitutive material of the composite structure, e.g., reinforced concrete structures contain triple face distributed springs for concrete material and only a single triple spring for each steel re-bar. These springs also allow for the implementation of the failure criteria properties associated with the structural component.

11.3 THE BRIDGE STRUCTURE

The analyzed bridge is the Yong-He Bridge, one of the first cable-stayed bridges constructed in mainland China and opened to traffic in December 1987. It has two main spans of 260 meters and two side spans of 25.15 and 99.85 meters each. The whole bridge is 510 meters long and 14.5 wide. The Yong-He Bridge is the object of an international benchmark proposed by the Centre for Structural Monitoring and Control at the Harbin Institute of Technology (Domaneschi et al. 2018, 2019) (Figure 11.1).

11.4 THE BRIDGE MODEL

The complete description of the finite element model of the bridge in the ANSYS code with the model updating procedure is presented in detail in the benchmark statement (see Domaneschi et al. 2018, 2019). The finite element model of the bridge has been validated on the basis of the field monitoring data from the full-scale bridge. It was developed on the basis of the engineering drawings and originally implemented using the ANSYS finite element (FE) software. The original benchmark model is consistently replicated using the AEM. The material parameters that were used in the original benchmark were implemented in the AEM model.

The AEM model has been validated through a comparison in terms of natural frequencies and mode shapes. The criterion used to verify the correlation between modes is the Modal Assurance Criterion (MAC). The MAC takes a value between 0 (representing no consistent correspondence) and 1 (representing a consistent correspondence). Values larger than 0.9 indicate consistent correspondence, whereas small values indicate poor resemblance of the two shapes. Seven shape modes are

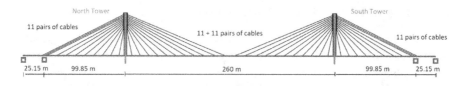

FIGURE 11.1 Yong-He Bridge geometry.

taken into account and all present MAC values greater than 0.9 (Domaneschi et al. 2018, 2019).

11.5 ROBUSTNESS AND REDUNDANCY ISSUES

In the literature, several competing approaches for the deterministic evaluation of structural robustness and redundancy have been presented. Among others, Biondini and Restelli (Domaneschi et al. 2018, 2019) evaluated robustness through an index that relates the global displacements of a structure composed of parallel members in different configurations: $\rho = \dfrac{S_0}{S_d}$.

Where S_0 is the displacement of the intact configuration of the system and S_d is the displacement of the damaged configuration. This robustness index ρ decreases from 1 to 0 as damage spreads within the system.

Liu et al. (Domaneschi et al. 2018, 2019) proposed three different indexes to assess redundancy components. Two indexes are related to the overloading of the originally intact configuration of the structure and are defined as the ability to withstand collapse and/or to avoid losses in the structural functionality. The third index is computed for a damaged configuration of the structure and allows the assessment of the system capability of carrying extra loads after the damage occurrence in the main structural member (Figure 11.2).

The redundancy indexes are defined in terms of the system reserve ratios such as R_u, R_f, R_d for the ultimate, functionality, and damaged condition limit states, respectively: $R_u = \dfrac{LF_u}{LF_1}$; $R_f = \dfrac{LF_f}{LF_1}$; $R_d = \dfrac{LF_d}{LF_1}$. Where LF_1 is the load that causes the failure of the first structural member, LF_u the load that is related to the achievement of the structural collapse, LF_f the load that induces the overcoming of the functionality limit state in the intact structure, LF_d the load connected to the collapse of the damaged structure (with one main member initially lost). In other words, R_u, R_f, R_d

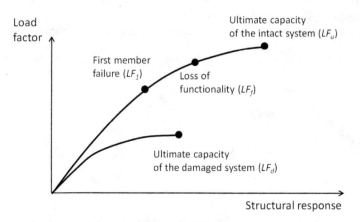

FIGURE 11.2 Load measures needed to calculate redundancy indexes.

indexes measure the system's capacity to withstand the first member failure and can be applicable for evaluating alternative design solutions. In Domaneschi et al. (2018, 2019), an algorithm to extend the methodology proposed by Liu et al. to a time history analysis is presented. Indeed, the original methodology was related to *pushover* analyses, therefore only the nonlinear static domain was considered.

11.6 SEISMIC RESPONSE SIMULATION OF THE YONG-HE BRIDGE: ORIGINAL AND MODIFIED CONFIGURATION

The effect of cable failure is investigated by nonlinear dynamic analyses of the time domain, taking into account large deformations of the original and a proposed modified configuration of the bridge (Domaneschi et al. 2018, 2019). The input accelerogram can analyze the influence of the phase-difference effect on the earthquake response of the Yong-He Bridge through the numerical simulations of this study. The proposed new configuration of the bridge implements a different arrangement for the stay-cables in the central span. The geometrical distance between the cables-deck connection is decreased while the cross-sections have been retained.

The results of the corresponding AEM analyses of the bridge highlight the disproportionate collapse shown in Figure 11.3. The new analyses that consider the new configuration of the deck supporting system show localized damage at the deck mid-span. Therefore the disproportionate collapse of the bridge has been avoided. The failure of the main girder is due to the high tensile stresses in the concrete (Domaneschi et al. 2018, 2019).

11.7 GENERAL REMARKS

Numerical analyses have the benefit of being able to effectively control the changes introduced in the modified structural configuration maintaining all other conditions unchanged. This allows a consistent comparison in terms of structural response between the original configuration and the alternatives.

The problem of double-cable-losses for the cable-stayed bridge is considered in Domaneschi (2018, 2019) at the numerical level with respect to the disproportionate collapse, and the way to improve redundancy is investigated through nonlinear

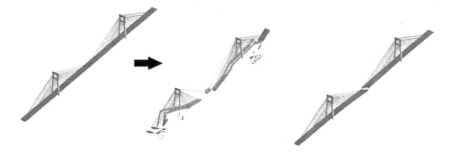

FIGURE 11.3 Disproportionate collapse from AEM simulations and modified bridge configuration.

time history analyses. It is worth noting how such a local failure condition improves the requirements of the existing PTI Recommendations. Indeed, they require that a cable-stayed bridge shall be capable of withstanding the loss of any one cable (PTI 2012). Therefore, the proposed alternative configuration of the cable-stayed bridge is able to give additional benefits with respect to the minimum expected requirements.

With respect to the existing literature, new redundancy indexes are proposed for time history analyses (Domaneschi 2018, 2019). They introduce new information, allowing a quantitative measure of redundancy. In particular, the proposed alternative bridge configuration, with respect to the original one, shows how the reduction of the geometrical distance between the cable-deck connection can uniformly improve all redundancy indexes by about 150%. This means that the system is capable of withstanding local cable failure with uniform reserve ratios, with respect to ultimate, functionality, and damaged condition limit states.

The improvements highlighted by the redundancy indexes between the alternative and the original bridge configurations are also reflected in the robustness index ρ, in agreement with the recognized relation between redundancy and robustness (Kanno and Ben-Haim 2011).

11.8 BENEFITS FROM STRUCTURAL CONTROL

Benefits can also be obtained from the implementation of intelligent control systems to compensate for any out-of-service structural elements through the parameter modifications of intelligent components. Figure 11.4 exemplifies the concept that has been termed in the literature as *Immediate Resilience*. When the shock hits the structure and causes partial out-of-service, the intelligent control system is able to compensate for part of the loss of functionality (red squared shape in Figure 11.4).

In Domaneschi and Martinelli (2016), the concept of seismic resilience of a controlled cable-stayed bridge was explored through investigations and finite element simulations on a case study. The case study is represented by a refined finite element model of an existent cable-stayed bridge, the object of an international benchmark (Figure 11.5).

The semi-active signature of the structural control elements is exploited online to compensate for losses of performance due to the failure of some of the control elements. Figure 11.5 also reports an example of out-of-service of two transversal

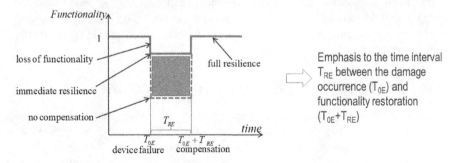

FIGURE 11.4 *Immediate Resilience* concept (Domaneschi and Martinelli 2016).

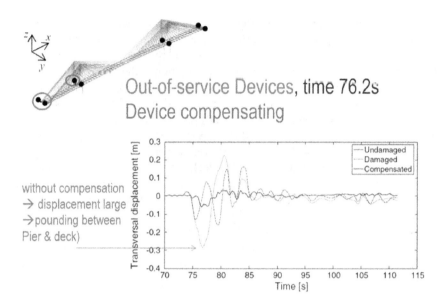

FIGURE 11.5 Example of application: *Immediate Resilience* for a cable-stayed bridge under seismic action.

devices at the bend (red circle). Automatically detecting the issue, in real-time with the transient response of the bridge, the device at the closest pier is able to modify its working parameters to compensate for possible performance amplification (red curve in the diagram of Figure 11.5). The compensated response (blue curve) is sufficient to avoid hammering phenomena between the deck and the pier, even if it essentially differs from the intact response of the original bridge configuration (black curve).

The proposed control solution's results show the ability to nullify the time interval between the damage occurrence and restoration (even if not complete) of the system's performance. This innovative aspect is related to how the devices' semi-active feature is exploited to enhance resilience. Therefore, the concept of *Immediate Resilience* is first introduced. A new measure of resilience is also proposed with reference to the performed simulations (Domaneschi and Martinelli 2016).

The positive outcomes coming from redundant and automatic seismic protection systems, such as the one here implemented, offer a contribution not only in undamaged working conditions but also when local failures occur, providing on-the-fly compensation to performance losses.

11.9 BENEFITS FROM STRUCTURAL HEALTH MONITORING

Monitoring is a central step for identification processes. With reference to damage detection activities they can be classified according to Doebling et al. (1996) into four sequential phases: recognition of the presence of the damage (*detection*), location of the damage (*localization*), *quantification* of the damage and finally *prognosis*, i.e., estimate of the remaining service life of the structure.

Monitoring systems may be useful to obtain information on the degradation conditions, to be able to adopt in advance the necessary actions, and, thus, reduce the risks of disproportionate collapses. In particular, a set of system identification techniques have been recently proposed using structural response variables only: they are usually mentioned as *output-only* techniques. These approaches originated with the need for operating without disrupting normal activities (e.g., traffic flow) or with the difficulty of consistently measuring the input loading (e.g., wind pressure, traffic load, etc.). Such innovative identification techniques can be successfully performed when the response of the structural system is independent of the input, or in other words when the transfer function of the system is independent of the external loading. It is usually related to stationary (or weakly stationary) white signals.

In Domaneschi et al. (2017), damage detection in composite concrete-steel structures, which are typical for highway overpasses and bridges, was investigated by using only structural response variables (the *output-only* technique). The method developed was based on the dynamic curvature analysis of real strain data from an in-service structure for real-time applications. A FE approach was also developed in parallel to interpret the structural behavior and then assess the effectiveness of the method. The data were acquired from long-gauge fiber optic strain sensors under the traffic loading of the structure. The probabilistic analysis of the peak values of dynamic curvature PSDs was used to study the real data and compare it to the FE results. The real data showed unusual behavior at one location where the average and standard deviation of the peak values of curvature PSDs were significantly higher than expected. Figure 11.6 summarizes the relevant characteristics of the study, where the unusual behavior in terms of the probabilistic response of the simply supported overpass is highlighted. These outcomes are in accordance with a previous study that identified unusual behavior at the same location.

In Domaneschi et al. (2016), an *output-only* arrangement of the interpolation damage detection method was checked for a suspension bridge based on responses

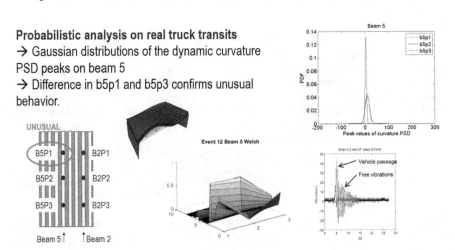

FIGURE 11.6 Output-only probabilistic technique for detection of unusual behavior on highway overpasses.

FIGURE 11.7 The Shimotsui-Seto (JP) Bridge and the FE model with wind forces.

to the wind-induced vibrations of a calibrated finite element model (Figure 11.7). Constant PSD of the input was assumed; even if not strictly verified, it is widely applied in output-only system identification. The effect of noise was evaluated for different damage intensities and positions with respect to a number of damage scenarios. The damage was modeled by a reduction, at several different positions, of the local stiffness in bridge deck members. Both noise-free and noise-polluted scenarios were considered in the numerical simulations. The *output-only* arrangement was demonstrated to be effective at the numerical level for Level II damage identification (damage localization) on a wind excited long-span bridge.

REFERENCES

Applied Science International, LLC (ASI) (2017), Extreme loading for structures Theoretical Manual – Extreme loading for structures 5.0, Applied Science International.

G.P. Cimellaro, A.M. Reinhorn, and M. Bruneau (2010), "Framework for analytical quantification of disaster resilience," *Engineering Structures*, 32(11), 3639–3649.

R. Das, A.D. Pandey, S. Soumya, and M.J. Mahesh (2016), "Assessment of disproportionate collapse behavior of cable stayed bridges," *Bridge Structures*, 12(1–2), 41–51.

S.W. Doebling, C.R. Farrar, M.B. Prime, and D.W. Shevitz (1996), Damage identification and health monitoring of structural and mechanical systems from changes in their vibration characteristics: A literature review, Report LA-13070-MS, Los Alamos National Laboratory.

M. Domaneschi, and L. Martinelli (2016), "Earthquake resilience-based control solutions for the extended benchmark cable-stayed bridge," *Journal of Structural Engineering, ASCE*, 142(8), art. no. 4015009.

M. Domaneschi, M.P. Limongelli, and L. Martinelli (2016), "Wind-driven damage localization on a suspension bridge," *The Baltic Journal of Road and Bridge Engineering*, XI(1), 11–21.

M. Domaneschi, D. Sigurdardottir, and B. Glišić (2017), "Damage detection based on output-only monitoring of dynamic curvature in concrete-steel composite bridge decks," *Structural Monitoring and Maintenance*, 4(1), 1–15.

M. Domaneschi, G. Scutiero, G.P. Cimellaro, A.A. Khalil, C. Pellecchia, and E.M. Ricciardi (2018), "Seismic resilience of bridges in transportation networks," *9th International Conference on Bridge Maintenance, Safety and Management (IABMAS 2018)*, Melbourne, Australia, July 9–13, 2018. ISBN-13: 978-1138730458, ISBN-10: 1138730459.

M. Domaneschi, G.P. Cimellaro, and G. Scutiero (2019), "Disproportionate collapse and disaster resilience of a cable-stayed bridge," *Bridge Engineering ICE*, 172(1), 13–26. doi:10.1680/jbren.18.00031.

Y. Kanno, and Y. Ben-Haim (2011), "Redundancy and robustness, or, when is redundancy redundant?," *Journal of Structural Engineering ASCE*, 137(9).

H.M. Salem, and H.M. Helmy (2014), "Numerical investigation of the collapse of the Minnesota I-35W bridge," *Engineering Structures*, 59, 635–645.

M. Shoghijavan, and U. Starossek (2018a), "Structural robustness of long-span cable-supported bridges segmented by zipper-stoppers to prevent progressive collapse," *IABSE Conference, Kuala Lumpur 2018: Engineering the Developing World – Report*, 593–600.

M. Shoghijavan, and U. Starossek (2018b), "Structural robustness of long-span cable-supported bridges in a cable-loss scenario," *Journal of Bridge Engineering ASCE*, 23(2), 04017133.

M. Shoghijavan, and U. Starossek (2018c), "An analytical study on the bending moment acting on the girder of a longspan," *Engineering Structures*, 167, 166–174.

X. Wang, B. Zhu, and S. Cui (2017), "Research on collapse process of cable-stayed bridges under strong seismic excitations," *Shock and Vibration*, art. no. 7185281, 18 pages.

M. Wolff, and U. Starossek (2009), "Cable loss and progressive collapse in cable-stayed bridges," *Bridge Structures*, 5(1), 17–28.

M. Wolff, and U. Starossek (2010), "Cable-loss analyses and collapse behavior of cable-stayed bridges Bridge Maintenance, Safety, Management and Life-Cycle Optimization," *Proceedings of the 5th International Conference on Bridge Maintenance, Safety and Management*, 2171–2178.

Z.H. Zong, X.Y. Huang, Y.L. Li, and Z.H. Xia (2016), "Study of collapse failure and failure control of long span cable-stayed bridges under strong earthquake excitation," *Bridge Construction*, 46(1), 24–29.

12 Building Resilience and Sustainability in Concrete Structures with FRP

Shamim A. Sheikh

CONTENTS

12.1 Introduction 189
12.2 Deficient Bridge Structure 190
12.3 Development of Rehabilitation Techniques 191
12.4 Bridge Repair... 193
12.5 Post-Repair Work.. 194
      12.5.1 Laboratory Experiments.. 194
      12.5.2 Field Test Data 196
12.6 Deficient Industrial Structure 198
12.7 Concluding Remarks 200
References.. 201

12.1 INTRODUCTION

Resilience of a city subjected to severe events, such as earthquakes or hurricanes, depends on many factors, an important one being the safety of its structures. The critical infrastructures such as hospitals, bridges, etc. must remain functional during an extreme event, and thus should be designed to withstand the forces and deformations imposed. But if a building in the vicinity of a hospital or a bridge collapses and blocks access to the critical structure, the planned strategy would not bear the expected results. There is a huge inventory of structures that are prone to collapse during a severe event. These structures are deficient either because of design flaws or due to deterioration as a result of aging. A major cause of this deterioration is the corrosion of steel, one of the most common building materials in the world. A 2010 study estimated that the total annual cost of corrosion from all sectors worldwide was $2.2 trillion, about 3% of the world's GDP of over $73 trillion (Hays 2010); approximately 16.4% of this total corrosion cost was from the infrastructure sector. Typically, corrosion in infrastructure occurs in bridge decks, beams, and columns leading to about $14 billion in annual direct cost for highway bridges in the United States alone (NACE International 2013). Figure 12.1 shows a bridge in Toronto in the 1990s before it was rehabilitated. Although originally it was designed and built properly, over time it potentially became a big hindrance to the resilience of the city, especially since a major hospital was within several meters of the structure.

FIGURE 12.1 Highway bridge damaged by steel.

In a complex structure, attention needs to be paid to the components that play a vital role in maintaining the integrity of the system. In this chapter, examples of such components are discussed, and solutions are suggested to address the deficiencies to ensure the survival and maintain the desired behavior through the extreme event. While the bridge in Figure 12.1 was damaged due to corrosion after decades of service, many existing structures are deficient as a result of design flaws or updated design codes. In some structures, some members may be shear critical, thus pointing to a brittle collapse during a severe earthquake. Research projects were undertaken to address these issues and selected results are briefly presented in this chapter.

12.2 DEFICIENT BRIDGE STRUCTURE

In the bridge shown in Figure 12.1, the damage was concentrated at the bents that are directly below the expansion joints. Over time, the joint sealant and other materials deteriorated due to extreme weather conditions and chemicals from the de-icing salts, allowing the water-ice-salts solution to pass through the joints to the reinforced concrete bents. Accumulation of this solution at the bent top and its continuous flow downward resulted in deterioration of concrete as well as corrosion of steel. The columns may also have been subjected to splashes of salt solutions from the cars using the area under the bridge as a parking lot. In addition to the direct chemical attack, the structure is also subjected to usual environmental effects that include freeze–thaw cycles, ultraviolet exposure, water, wet–dry cycles, etc. The spiral steel in the columns and tie steel in the beams were severely corroded and as a result the cover concrete in both columns and girders was severely delaminated.

Conventional repair techniques are cumbersome and usually require the closing of the entire facility during repair. They would involve providing temporary supports to the bridge while the contaminated concrete and corroded steel are removed. The structure would then be rebuilt with new steel and cementitious materials, including concrete. This procedure is costly and generally would not last for more than ten years before requiring a repetition of the repair. A rehabilitation technique was developed which allowed regular activities to continue with minimal disruption and provided a durable solution not requiring a repeat of the repair. The cost was also substantially lower than that for traditional methods.

12.3 DEVELOPMENT OF REHABILITATION TECHNIQUES

In the research program undertaken to address the issue discussed above, ten approximately half-scale models, 406 millimeters in diameter and 1.37 meters long, of the bridge columns were constructed; five were intended for short-term testing and the other five for a long-term investigation (Sheikh, S. A.). Each column was reinforced with 6–20M longitudinal reinforcing bars and a 10M spiral with 75 millimeters pitch. After 34 days, eight of these specimens were subjected to an accelerated corrosion process to produce damage similar to that observed in the field. Six corroded column specimens were repaired using different retrofit procedures. Two of the columns damaged by corrosion were not repaired and were used as control specimens along with two un-corroded healthy columns. Figure 12.2 shows a column with simulated damage and two columns in different stages of repair. The repair work was aimed at minimizing the time required to repair and its cost. It was decided that the corroded steel and the contaminated concrete would not be removed from the damaged columns. It was also decided to utilize fiber-reinforced polymers (FRP) for the upgrade and investigate the use of a plastic sheet wrapped around the

FIGURE 12.2 Damaged column and repair steps in the laboratory.

column as a formwork for grout – this acted as a barrier between the new grout and glass FRP (GFRP). The barrier between GFRP and the fresh grout was considered necessary since simulated lab studies on fibers immersed in sodium hydroxide solution had reported adverse effects of alkalis on the mechanical properties of glass fibers (e.g., Uomoto and Nishimura 1999). The results from the short-term tests on four columns from this study are presented here in Figure 12.3.

The Emaco-repaired column in Figure 12.3 was built to its original shape with the commercially available rheoplastic, shrinkage-compensated mortar called Emaco. The new grout was covered with a protective epoxy coating to avoid direct contact between the new cementitious material and GFRP. After 24 hours of curing, the column was wrapped with two layers of GFRP. One column in Figure 12.3 was repaired with grout based on the expansive cement (Exp.-repaired) developed as part of our research to be used in closed spaces for self-stressing (confinement) of concrete. A 3-millimeter thick polymer sheet reinforced with polyethylene fibers was wrapped around the damaged area of the column and held in place with five hose clamps so as to act as formwork for the process of column repair. Next, the expansive cement mortar was poured in place. Four hours after grouting, the column was wrapped with two layers of GFRP on top of the 3-millimeter thick polymer formwork sheet. After repair, the columns were stored in a lab at a temperature of 23°C and a relative humidity of about 50% for a duration of three months. Following this, they were tested for short-term behavior.

It can be seen from Figure 12.3 that the damaged column had about 20% lower strength due to corrosion of steel and loss of concrete cover compared with the undamaged control column specimen. Reduction in ductility and energy dissipating capacity was observed to be even higher. As shown in Figure 12.2, the cover concrete had almost entirely spalled off or become ineffective in all the corroded columns. Figure 12.4 shows four column specimens at the end of the tests. It can be seen that corrosion rendered the spiral steel completely ineffective in the damaged columns; pitting corrosion had created weak locations along the spirals, which resulted

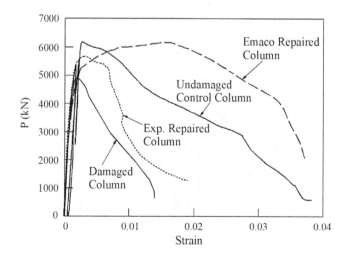

FIGURE 12.3 Behavior of columns under axial load.

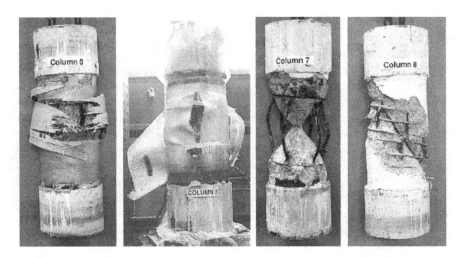

FIGURE 12.4 Columns after testing.

in them rupturing at a relatively small axial strain. Thus, the confinement that is needed for large strains is not available in corroded columns. These columns make the structure susceptible to collapse under any extreme event or even under regular usage, i.e., gravity loads alone. Consequently, a resilient structure ends up becoming non-resilient as a result of the corrosion-induced aging process.

Both repair methods displayed in Figure 12.3 improved the behavior of the specimens relative to the damaged unrepaired specimen. The axial load carrying capacity of the Emaco-repaired specimen was as large as the control undamaged column specimen. Its ductility and energy dissipation capacity were substantially better and exceeded the ductility properties of the undamaged control column. The Exp.-repaired column, in which plastic sheet formwork was used, did not display as good a performance as the Emaco-repaired column, in which no formwork was used. At large strains, the plastic sheet of the formwork opened and engaged the fiber wrap, creating large local strain and premature failure of the wrap, and hence the column. Figure 12.4 clearly shows the opening of the formwork sheet in the Exp.-repaired column and its adverse effects on FRP in comparison with the Emaco-repaired column. It was thus decided not to use the plastic sheet in the field and instead employ removable steel forms for Exp.-repaired columns. The separation between the new cementitious grout and GFRP was achieved with a thin polyethylene sheet

12.4 BRIDGE REPAIR

All the columns and beams in the bridge were repaired using the techniques developed in this research. Three of the columns shown in Figure 12.1 were repaired using three different repair schemes and monitored to evaluate the long-term performance of the techniques developed.

Column 124-1 was repaired with expansive concrete grout using steel formwork. The grout, which replaced the original cover concrete, was about 50 millimeters in thickness. About 20 hours after grouting, the formwork was removed, and the column was wrapped with a thin polyethylene sheet followed by two layers of GFRP with the glass fibers aligned in the circumferential direction only. The polyethylene sheet acted as a barrier between the new concrete and the glass FRP and was not expected to affect the column behavior under load. The bond between the GFRP and the concrete was not expected to play any significant role in providing confinement to the columns. Three days after the grout application, the column was instrumented with six strain gauges in the circumferential direction installed on the FRP. Two gauges each, 180° apart, were installed at mid-height, 750 millimeters above and 750 millimeters below the column mid-point.

Column 124-2 was repaired using a commercially available non-shrink grout, which was pumped in place with the steel formwork. Four days later, the steel forms were removed and the column was wrapped in the polyethylene sheet and GFRP in the same way as Column 124-1. Eight days after the application of grout, this column was instrumented in a manner similar to Column 124-1.

Column 124-3 was built to its original shape with EMACO-based mortar. No formwork was needed for this rheoplastic mortar. Six days later, a protective epoxy coating was applied over which two layers of GFRP were installed. The instrumentation in this column was similar to Column 124-1. The gauges were applied to this column eight days after grout application.

To monitor corrosion activity in the repaired columns, three half-cells (Silver/Silver Chloride) were embedded in each of the three field test columns. The cells were located at the top, middle, and bottom of the columns. The corrosion potential from these cells was measured in mV at regular intervals for more than eight years.

12.5 POST-REPAIR WORK

Two phases of post-field-repair work were undertaken to evaluate the long-term performance of the structure for resilience. The field monitoring was carried out to monitor the long-term durability of the rehabilitation techniques under natural environment. The laboratory tests were undertaken to determine the contribution of FRP toward improving the mechanical resistance of structures against extreme events.

12.5.1 Laboratory Experiments

Hundreds of specimens were tested for material characteristics and structural behavior investigating a variety of variables, such as column size and shape, type of FRP, concrete properties, environmental conditions, types of loads, etc.

In the following sections, results from a few specimens are shown in which the columns were subjected to simulated seismic loads. The focus of these tests was to quantify the beneficial effects of FRP wraps. All the columns were 356 millimeters in diameter, 1.47 meters long, and cast integrally with a stub of dimensions 510 × 760 × 810 millimeters representing a beam-column joint or a foundation. The column portion of the specimen represented the part of a column in a bridge between the section of maximum moment and the point of contra-flexure. Details of the tests and testing procedure can be found elsewhere (e.g., Sheikh and Yau 2002; Kharal and Sheikh 2018).

All columns contained six 25M longitudinal bars uniformly distributed around the core. The lateral reinforcement of #3 US spiral at 300 millimeters represented deficient confinement for seismic resistance either through design flaws or as a result of corrosion of steel. The moment-curvature responses of the critical sections are shown in Figure 12.5 for column specimens S-2NT, S-4NT, and ST-5NT. All the columns were tested under an axial load of 0.27 P_o, where P_o represents the theoretical

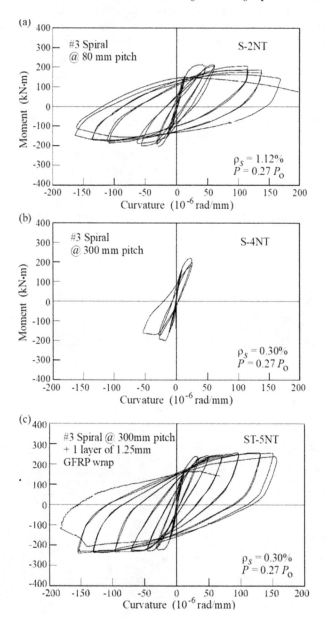

FIGURE 12.5 Moment-curvature responses of columns.

axial load carrying capacity of the column. Specimen S-2NT contained spiral steel at a 75-millimeter spacing thus representing a column that is healthy and designed in accordance with the design codes (CSA 2004).

Specimen S-4NT contained spiral steel at a 300-millimeter spacing that represented a column in which spiral steel was lost due to corrosion or the column was built with a deficiency in transverse reinforcement. The presence of a sound concrete cover in Column S-4NT would overestimate the response of a column that is damaged by corrosion and has lost the concrete cover but only at small deformations. Column ST-5NT contained spiral steel at a 300-millimeter spacing and was retrofitted with one layer of GFRP wrap. This specimen thus represented a repaired column in which the spiral steel was not replaced after damage. A comparison of the moment-curvature responses of Specimens S-2NT and S-4NT underlines the importance of the spiral steel and its spacing on the behavior of the columns. Strength, ductility, energy dissipation capacity, and the number of cycles of inelastic excursions that a column can sustain depends on the effectiveness of the spiral steel and the confinement it provides. The deficiency of the column as a result of less than adequate spiral steel can be easily overcome by the addition of only one layer of GFRP wrap, as shown by the response of Specimen ST-5NT. Strength, ductility, and the energy dissipation capacity of GFRP-retrofitted Column ST-5NT are similar to or superior to those of the control Column S-4NT that meets the current seismic code provisions.

12.5.2 Field Test Data

Lateral strain data collected from the field columns have been plotted against time in Figure 12.6. As expected, Column 124-1 displayed substantial expansion with FRP

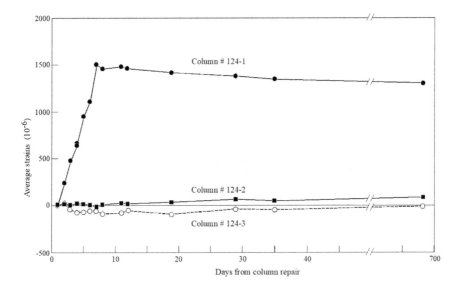

FIGURE 12.6 Lateral strain in GFRP in repaired columns.

Building Resilient and Sustainable FRP Concrete Structures

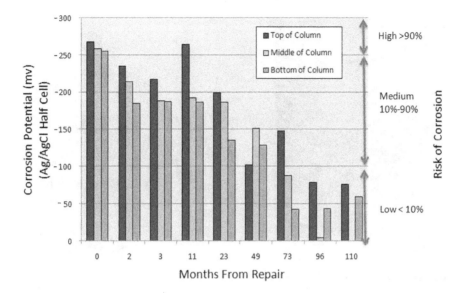

FIGURE 12.7 Corrosion activity in repaired columns.

strain of about 0.0015, which is less than 10% of the GFRP rupture strain. It should be noted that the normal yield strain for steel used as transverse reinforcement is approximately 0.002. No significant lateral strain was measured in GFRP in the other two columns. Lateral GFRP strain in all the columns remained fairly constant for about two years, indicating no significant creep of GFRP. Recording of strain data was terminated about two years after repair.

Figure 12.7 shows the corrosion potential and risk of corrosion in one of the repaired field columns at different locations along its height for over nine years. This data is based on the measurements from half-cells embedded in the columns. Soon after the repair in 1996, based on the average of potential measurements at three locations along the height of each column, the risk of corrosion in the repaired Columns 124-1 and 124-2 was found to be high, and in Column 124-3 it was intermediate. Eight years later, the risk of corrosion in all the columns was low. Reduction in the corrosion activity and reduced risk of corrosion can be clearly seen in Figure 12.7. It is clear that GFRP wraps protected the columns from adverse environmental effects, thus starving the corrosion of its essential ingredients, water and oxygen, resulting in reduced corrosion activity. As a result, the repair proved to be durable.

Monitoring of the columns over the years through visual inspection and field data on strain and corrosion rate indicated a sound performance of the retrofit techniques. No distress or deterioration was observed in these columns after about 20 years of service, as shown in Figure 12.8. In the repair of the columns, a thin polyethylene sheet was used as a barrier to separate new mortar or concrete from the GFRP to avoid any possible adverse effects of alkalis on the performance of GFRP in the long term. Since then, extensive testing has shown the excellent long-term performance of GFRP sheets under an alkaline environment (Homam et al. 2001; Sheikh and

 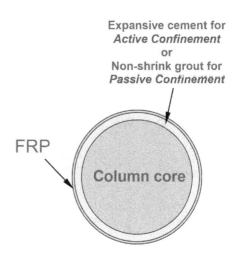

FIGURE 12.8 Rehabilitated columns after twenty years of service.

Homam 2004). Repairs using GFRP are now being carried out without the use of barriers. Although the presence of the barrier is not necessary for isolating glass from alkali, the barrier might have been partially responsible for the reduced corrosion activity in the field columns studied.

12.6 DEFICIENT INDUSTRIAL STRUCTURE

Figure 12.9 shows part of a seven-story industrial building from a cement manufacturing plant. The height of the structure is about 81 meters. A detailed analysis of the structure showed that some of the beams were shear critical. During a seismic event, the collapse of the frame would occur due to shear failure of beams although columns were well designed and would not fail. A half-scale two-dimensional model of the two stories of the structure was built and tested in the structures laboratories at the University of Toronto in two phases (Duong et al. 2007). In Phase A, the beams were severely damaged in shear. For Phase B, they were repaired with carbon FRP. Figure 12.10 shows pictures of the two phases of the frame's test.

The load-deflection responses of the two phases of the test are compared in Figure 12.11. It is clear that the failure of the beams in shear in Phase 1 resulted in a brittle behavior of the frame resulting in impending collapse. The beams were retrofitted with one layer of carbon FRP in shear after filling the cracks with epoxy. No other modifications were made to the frame. The test was then repeated under lateral displacement excursions and terminated only when the maximum displacement capacity of the actuator was reached. The maximum displacement measured of the upgraded frame was about four times that of the original frame, and the energy dissipation increased by a factor of about six. The response of the frame in Phase 2 shows a remarkable improvement over Phase 1 resulting in large increases in ductility, energy dissipation capacity and robustness of the structure.

Building Resilient and Sustainable FRP Concrete Structures

FIGURE 12.9 Deficient industrial structure.

Phase A: Initial Condition   Phase A: Final Condition   Phase B: Initial Condition   Phase B: Final Condition

Phase A, Original frame   Phase B, Carbon FRP- retrofitted beams

FIGURE 12.10 Model frame test with shear critical beams. Phase A: Original frame; Phase B: Carbon FRP-retrofitted beams.

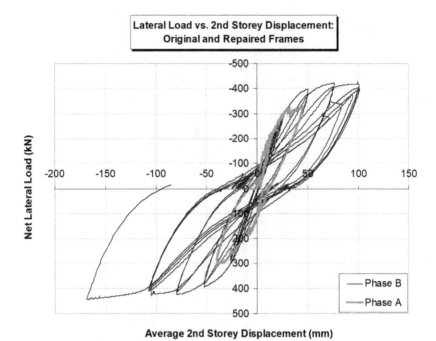

FIGURE 12.11 Load-displacement response of frame.

12.7 CONCLUDING REMARKS

Several bridge columns and beams damaged by steel corrosion along a major highway in Toronto made the bridge seriously deficient. An industrial structure was also studied, which was found to be shear critical and deficient for seismic resistance. Based on an extensive research program in which half-scale models of the prototypes were tested in the lab, innovative techniques were developed to rehabilitate the structures in a cost-effective manner with minimal closure time of the structures. The non-traditional materials used in the upgrade/repair included especially developed expansive cement and glass and carbon-fiber-reinforced polymers. The bridge structure was closely monitored for over ten years after rehabilitation especially for corrosion. Although the corroded steel and the contaminated concrete were not removed from the structure, field measurements showed that the corrosion activity and risk of corrosion reduced with time in the repaired columns. The upgraded industrial structure has withstood severe earthquakes without any serious damage.

Based on the laboratory studies and field monitoring, it is concluded that innovative solutions involving FRP and specialized grouts can help upgrade structures for sustainability under severe load and environmental conditions. The upgraded structures have performed flawlessly for over 20 years, indicating high durability. It can be concluded that resilience and sustainability can be built into the deficient or damaged structures through innovative durable techniques employing new materials such as fiber-reinforced polymers (FRPs).

REFERENCES

Canadian Standards Association, *Design of Concrete Structures, CSA-A23.3-04*, Rexdale, Ontario, 2004.

Duong, Kien Vinh, Sheikh, Shamim A. and Vecchio, Frank, "Seismic behavior of a shear-critical reinforced concrete frame: Experimental investigation," *ACI Structural Journal*, 104(3), 2007, pp 304–313.

Hays, G. F., "NACE-International, the corrosion society" 2010. http://impact.nace.org/economic-impact.aspx).

Homam, S. M., Sheikh, S. A. and Mukherjee, P. K., Durability of fibre reinforced polymers (FRP) wraps and external FRP-concrete bond. *Proceedings of the Third International Conference on Concrete Under Severe Conditions*, Vancouver, BC, Canada, 2, pp. 1866–1873, 2001.

Kharal, Z. and Sheikh, S. A., "Seismic performance of square concrete columns confined with glass fiber–reinforced polymer ties," *ASCE Journal of Composites for Construction*, 22(6), December 2018. https://ascelibrary.org/doi/full/10.1061/(ASCE)CC.1943-5614.0000884.

NACE International, "Corrosion central, industries and technologies, highways and bridges", 2013. https://store.nace.org/nace-publication-10a292-2013-2.

Sheikh, S. A., "Field and lab performance of bridge columns repaired with wrapped GFRP," *Canadian Journal of Civil Engineering*, National Research Council, Canada, 34(2), pp 403–413, 2007.

Sheikh, S. A. and Homam, S. M., A decade of performance of FRP-repaired concrete structures. *Proceedings of the 2nd International Workshop on Structural Health Monitoring of Innovative Civil Engineering Structures*, Winnipeg, Manitoba, pp. 525–534, 2004.

Sheikh, S. A. and Yau, G., "Seismic behavior of concrete columns confined with steel and fiber-reinforced polymers," *ACI Structural Journal*, 99(1), pp. 72–80, 2002.

Uomoto, T. and Nishimura, T., Deterioration of aramid, glass, and carbon fibres due to alkali, acid, and water in different temperatures. *Fourth International Symposium on FRP Reinforcement for Concrete Structures*, Dolan, Rizkallah, and Nanni (Eds.), Farmington Hills, Michigan, American Concrete Institute, pp. 515–522, 1999.

13 Resilience-Oriented Displacement-Based Seismic Design Procedure and Its Application in Self-Centering Metallic Energy-Dissipating Structures

Lu Liu and Bin Wu

CONTENTS

13.1 SC-MED Structures 203
13.2 The R-C_R DDBD 204
13.3 Application of R-C_R DDBD in the SC-MED Structures 205
13.4 Conclusion 210
References 211

13.1 SC-MED STRUCTURES

A self-centering (SC) structure is a high-performance lateral-resistance system, wherein residual deformation can be well controlled, if properly designed. Various energy-dissipating (ED) devices are incorporated in the SC structures to reduce maximum deformation. One such ED device is the metallic energy-dissipating (MED) device (Tena-Colunga et al., 2019), with which the hysteretic behavior can usually be conservatively simplified into elastoplastic hysteresis, as illustrated in Figure 13.1 (d).

The SC-MED structure is conceptually divided into the SC system and MED devices (Figure 13.1 (a)). Since the frame members are designed to remain elastic under a prescribed earthquake, the SC system is essentially a superposition of the SC device and the elastic frame members and, thus, is featured by bilinear elastic hysteresis, as shown in Figure 13.1 (c). As a combination, the SC system and the MED

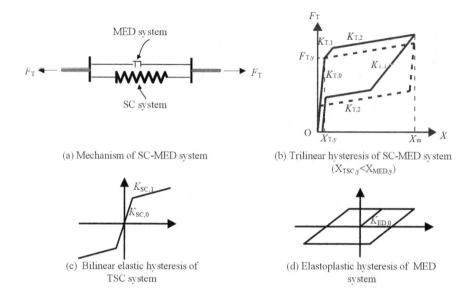

FIGURE 13.1 Concept and restoring force of SC-MED structure. (From Liu et al, 2018a.)

devices, the hysteresis of an SC-MED system is usually represented by the solid line in Figure 13.1 (b), featured by trilinear hysteresis. However, in previous studies on the seismic response of SC-MED systems, the bilinear hysteresis as given by the dashed line in Figure 13.1 (b) is widely used, such as in SC buckling-restrained braced frame (SC-BRBF) (Liu, 2012; Liu et al., 2018a; Liu et al., 2018b; Chou et al., 2014; Zhou et al., 2015), SC-MED MRF (Ricles et al., 2001), and SC-MED base-rocking systems (Perez et al., 2007; Martin et al., 2019; Lu et al., 2017). It is found that such approximation could largely underestimate the nonlinear displacement demand, and thus is non-conservative. Therefore, Liu et al. proposed a practical equation for evaluation of the nonlinear displacement of SC-BRBF, considering the trilinear hysteresis of SC-BRBF (Liu et al., 2018a). Furthermore, the result could be extended to other SC-MED systems.

Due to great seismic-resistance advantages, several SC-MED systems have been constructed in New Zealand. However, "the SC-MED structure was so groundbreaking that the calculations, testing, and performance reviews took place before the guidelines for the new technology had even been completed." Therefore, a design method is required for this innovative structure.

13.2 THE R-C_R DDBD

The performance-based seismic design (PBSD) procedure has been codified in many provisions. In PBSD, the structures are designed to satisfy different performance objectives of the corresponding seismic level. In the framework of PBSD, the displacement-based design (DBD) principle takes the displacement indexes as engineering demand parameters (EDP), so that the damage of some ductile structure elements can be well-reflected. There are mainly two branches of DBD, the $R_\mu - \mu - T$-based procedure (Miranda and Bertero, 1994; Vidic et al., 2010; Xue

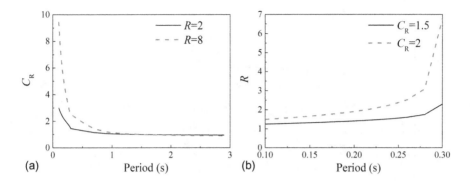

FIGURE 13.2 Plot of $C_R - R(-T)$ and $R - C_R(-T)$ relation. (From Liu et al, 2018b.)

and Chen, 2003; Qiu and Zhu, 2017) and the ELM-based procedure (Priestley, 1993; Suarez and Kowalsky, 2010; Roldan et al., 2016).

To improve the current DBD, a direct displacement-based design (DDBD) procedure without iteration, which is more accurate regardless of the seismic level, is proposed (Liu, 2012; Liu et al., 2018b), and applied to the SC-MED systems, on the basis of the inelastic deformation ratio. The inelastic deformation ratio C is defined as the ratio between the maximum displacement X_{non} of a nonlinear system and displacement X_e of a linear system with the same initial period, subjected to a specific seismic record. By calculating the inelastic deformation ratio of a number of SDOF systems with various periods and constant R, the C_R spectra can be obtained, i.e., the C_R-R relation, as illustrated in Figure 13.2 (a). As the inverse function of the C_R-R relation, the R-C_R relation as illustrated in Figure 13.2 (b), which is defined as the function of strength reduction factor R with respect to nonlinear displacement ratio C_R, is the essence of the innovative R-C_R DDBD presented (Liu, 2012; Liu et al., 2018b). The R-C_R relation of the SC-MED structures is given by

$$R = f(T, C_R, \alpha_c, \alpha_s, \beta) \tag{13.1}$$

Where α_c, β, and α_s are normalized hysteretic parameters, which define a group of hysteresis of the same initial stiffness K_0, with geometric similarity, as illustrated in Figure 13.3. α_c is defined as $K_{ED,0}/K_{SC,0}$, and α_s is defined as $K_{SC,1}/K_{SC,0}$; β is the ratio of the yield strength of the MED system to that of the TSC system (Liu, 2012; Liu et al., 2018b), where the connotations of $K_{ED,0}$, $K_{SC,0}$, and K_0 are illustrated in Figure 13.1 (c) and (d).

13.3 APPLICATION OF R-C_R DDBD IN THE SC-MED STRUCTURES

The R-C_R DDBD, which is normally implemented in two steps (the linear design and the nonlinear design), is validated herein in a three-story planar SC-BRBF (elevation illustrated in Figure 13.4) subjected to MCE earthquakes. The foundation of the SC-BRBF is fixed, and beams are pinned to the columns. The same frame members and braces are assigned in each story of the building. The input parameters of the DDBD are the seismic level and the objective roof drift ratio [RDR]. In the first place,

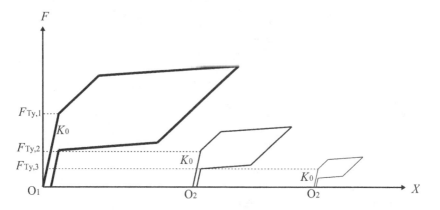

FIGURE 13.3 Hysteresis of SC-MED structures with the same initial stiffness K_0, α_c, β and α_s, and different yield strength, $F_{Ty,1} = 2F_{Ty,2} = 4F_{Ty,3}$.

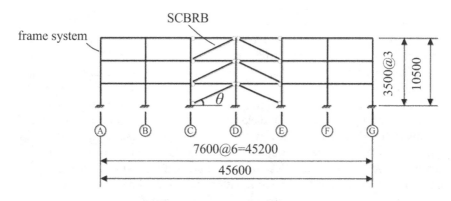

Columns Ⓐ Ⓑ Ⓕ Ⓖ: GB-HM350X350x12x19
Columns Ⓒ Ⓓ Ⓔ : ☐400x400x25x25
Beams Ⓒ-Ⓔ: GB-HW400x400x13x21
Beams Ⓐ-Ⓒ, Ⓔ-Ⓖ: GB-HM350X350x12x19

FIGURE 13.4 Elevation of the SC-BRBF.

the dimensionless key hysteresis parameters are specified as $\alpha_s = 0.3$, $\alpha_c = 0.5$, and $\beta = 0.7$, and assumed to be unchanged for each story, as recommended (Liu, 2012; Liu et al., 2018b). In addition, the stiffness proportion $\alpha_{B/M}$ between the SC-BRB and the frame members is specified as 6, according to the objective drift ratio and type of beam-to-column connections, as recommended in (Liu, 2012; Liu et al., 2018b).

$$\alpha_{B/M} = K_B/K_M = 2k_{b,i}/K_{M,i} \qquad (13.2)$$

Where K_B and K_M denote the initial lateral stiffness of the brace and frame system, respectively; $k_{b,i}$ is the initial lateral stiffness of an SC-BRB at the ith story; $K_{M,i}$ is the frame stiffness at the ith story.

At the end of the linear design, the initial stiffness of all frame members and braces are obtained. The fundamental period T_0 and T_M of the SC-BRBF and the frame system are evaluated using Equation (13.3) and Equation (13.4), respectively.

$$T_0 = 0.1N = 0.3 \text{ s} \tag{13.3}$$

$$T_{M0} = T_0\sqrt{(1+\alpha_{B/M})} = 0.79 \text{ s} \tag{13.4}$$

The cross-sections of the frame members are given in Figure 13.4, with $T_M = 0.79$ s. The columns in the braced bays are designed with extra member size to withstand the axial force exerted by the braces under earthquakes. Using Equation (13.2), the initial lateral stiffness $k_{B0,I} = 250$ kN/mm of an SC-BRB is determined. Actual $K_{M,I} = 85.7$ kN/mm, $\alpha_{B/M} = 5.8$, and $T_0 = 0.27$ s of the designed SC-BRBF are determined using linear pushover analysis. Note that the selection of linear system is only one choice among many, with the intention that the frame members of the selected SC-BRBF do not yield under the prescribed objective drift ratio in the pushover analysis.

Then follows the nonlinear pushover analysis of SC-BRBF with elastic braces, to determine the yield drift ratio of the system (0.7%). A similar analysis is performed for the frame system without braces, and a yield drift of 1.67% is determined. Therefore, it could be inferred that the yield drift of the designed building lies between 0.7% and 1.67%, so the objective drift should be selected in this scope to ensure that the frame members remain elastic.

Three seismic records (RSN14_KERN_SBA042, RSN15_KERN_TAF021, RSN33_PARKF_TMB205) are selected to fit the mean truncated DBE spectrum of LA (Figure 13.5), of which $S_{ds} = 1.53$ g, $S_{dl} = 0.6997$ g, and $T_L = 8$ s are stipulated in ASCE 7-10. Good fitness of the spectral acceleration is observed in the period range from 0.24 s to 0.28 s, which covers the fundamental period of the SC-MEDF. The MCE acceleration records are obtained by multiplying the three DBD acceleration

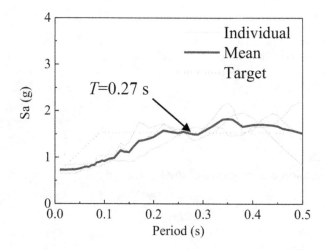

FIGURE 13.5 Truncated elastic acceleartion DBE spectra for LA.

records by 1.5. The linear demand under MCE level earthquakes determined using linear time history analysis, instead of response spectra analysis, is listed in Table 13.1. Specifically, the mean roof drift $X_e = 36$ mm, roof drift ratio $RDR_e = 0.36\%$, and mean base shear demand $F_e = 8506$ kN are determined.

In the nonlinear design phase, the period of the equivalent SDOF of the SC-BRBF, $T_e = 0.27$ s is calculated using the procedure proposed by Fajfar (2000). The calculating process is illustrated in Table 13.2, where ϕ_i and m_i are the component of the first modal vector corresponding to the ith floor and story mass at the ith floor, respectively. PF is the modal participation factor and M_1^* is the first modal mass. K_i is the identical story stiffness.

An objective $[RDR] = 1\%$ is specified to ensure elastic frame members, according to the result of pushover analysis. Thus, the objective nonlinear displacement ratio, $[C_R] = 3.33$, is determined by

$$[C_R] = [RDR]/RDR_e \qquad (13.5)$$

TABLE 13.1
Linear and Nonlinear Demand of the Designed SC-BRBF

EDP	SBA042	TAF021	TMB205	Mean
Linear roof drift X_e (mm)	42	38	28	36
Linear inter-story drift ratio at the first story, $IDR_{e,1}$ (%)	0.39	0.36	0.26	0.34
Linear inter-story drift ratio at the second story, $IDR_{e,2}$ (%)	0.44	0.41	0.3	0.38
Linear inter-story drift ratio at the third story, $IDR_{e,3}$ (%)	0.26	0.25	0.18	0.23
Linear roof drift ratio, RDR_e	0.4%	0.36%	0.27%	0.36%
Linear shear at the first story (base shear), $F_{e,1}$ (kN)	9894	9022	6602	8506
Linear shear at the second story, $F_{e,2}$ (kN)	7779	7163	5259	6734
Linear shear at the third story, $F_{e,3}$ (kN)	4345	4039	2973	3786
F_e (kN)	9894	9022	6602	8506
X_{non} (mm)	155	93	56	101
Nonlinear inter-story drift ratio at the first story, $IDR_{non,1}$ (%)	1.17	0.72	1.3	1.06
Nonlinear inter-story drift ratio at the second story, $IDR_{non,2}$ (%)	2.0	1.2	1.2	1.47
Nonlinear inter-story drift ratio at the third story, $IDR_{non,3}$ (%)	1.34	0.9	0.7	0.98
Nonlinear roof drift ratio, RDR_{non}	1.49%	0.89%	0.53%	0.965%

TABLE 13.2
Calculation of the Period of the Equivalent SDOF

Floor	ϕ_i	m_i (t)	$m_i\phi_i$	$m_i\phi_i^2$	PF	M_1^* (t)	K_i (kN/mm)	T_e (s)
3	1	228	228	228	1.2	611	217	0.27
2	0.77	228	176	135.9				
1	0.40	228	90	35.8				

It is pointed out by Wiebe and Christopoulos that SC structures could be designed with values of R more than 20 (Wiebe and Christopoulos, 2014; Steele and Wiebe, 2016), indicating R values greater than 8 are permissible in the SC structures. In our recent study, the values of R no more than 14 were stipulated for the SC-MED structures (Liu et al., 2019). However, the determined objective strength reduction factor $[R]$, is much greater than the upper bound of R, i.e., $R = 14$. Therefore, $[C_R]$ should be recalculated using $R = 14$. Table 13.3 shows parameter combinations around the period of 0.27 s (Table 13.3), truncated from massive regression data for Equation (13.1), with R values from 10 to 14. According to Table 13.3, with R values no more than 14, we can only obtain an SC-MED structure with an uplimit $[C_R] = 2.67$, which is less than the $[C_R]$ determined by Equation (13.5). Consequently, $[RDR] = 2.67 \times 0.36\% = 0.96\%$ is adopted as the final objective nonlinear drift ratio. The yield base shear is 8506 kN/14 = 614 kN, identical to the yield strength at each story.

Based on these, the remaining brace parameters of each SC-BRB are determined as follows (Table 13.4). The initial stiffness of the brace ED system, $k_{ED,0} = 118$ kN/mm, the initial and second stiffness of the brace SC system, $k_{SC,0} = 185$ kN/mm and $k_{SC,1} = 19.7$ kN/mm, the yield strength of the brace ED system, $f_{ED,y} = 158$ kN, the yield strength of brace SC system, $f_{ED,y} = 176$ kN.

The nonlinear response of the designed SC-BRBF is investigated and the EDPs under each record are listed in Table 13.4. It should be noted that the record-to-record (RTR) dispersion is evident in the nonlinear demand, which is also observed in the SDOF (Liu, 2012; Liu et al., 2018b). The nonlinear roof drift ratio RDR_{non} under record SBA042 is 1.49%, and plastic hinges corresponding to life safety

TABLE 13.3
Parameter Combination of Interest for SDOF SC-BRBF

T (s)	α_c	α_s	R	β	C_R
0.26	0.5	0.3	10	0.7	2.44
0.28	0.5	0.3	10	0.7	2.21
0.26	0.5	0.3	12	0.7	2.57
0.28	0.5	0.3	12	0.7	2.33
0.26	0.5	0.3	14	0.7	2.67
0.28	0.5	0.3	14	0.7	2.41

TABLE 13.4
Key Parameters of SC-BRBs at Each Floor

floor	$k_{SC,0}$ (kN/mm)	$k_{SC,1}$ (kN/mm)	$f_{SC,y}$ (kN/mm)	$k_{ED,0}$ (kN/mm)	$f_{ED,y}$ (kN/mm)
1	185	19.7	176	118	158
2	185	19.7	138	118	124
3	185	19.7	77	118	69

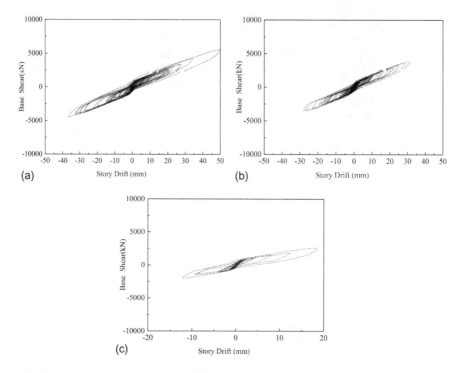

FIGURE 13.6 Dynamic hysteresis of the first story.

(LS) level appear in this case. However, the mean drift from nonlinear time history analysis under the three records is 0.965%, quite close to the objective 0.96%, so the design result of this case is satisfactory in terms of the mean trend. The hysteresis behavior of the first story subjected to the three records is illustrated in Figure 13.6.

Also listed in Table 13.1 are the inter-story drift ratio s(IDR) at each story, where $IDR_{e,1}$, $IDR_{e,2}$, and $IDR_{e,3}$ are the linear inter-story responses, and $IDR_{non,1}$, $IDR_{non,2}$, and $IDR_{non,3}$ are the nonlinear linear inter-story drift ratios. According to Table 13.4, the maximum nonlinear IDR equal to 1.47% occurs on the second floor. Since the objective nonlinear inter-story drift ratio $[IDR]_{non}$ is evaluated as 1.33–1.55 times the objective nonlinear roof drift ratio $[RDR]_{non}$ in a properly-designed SC structure (Qiu and Zhu, 2016), an $[IDR]$ = 1.33–1.49% is predicted. Therefore, the design is satisfactory in terms of the inter-story drift.

13.4 CONCLUSION

A resilience-oriented displacement-based design (R-C_R DDBD) procedure previously-proposed is used for the design of the SC-BRB steel frame, as an example. In this procedure, the yield base shear of the structure is directly determined by the objective nonlinear displacement ratio. Nonlinear pushover analysis is performed to identify the scope of the yield drift of the framing system for the selection of objective drift ratio and to ensure elastic frame members. The design result of the example considered is quite satisfactory in terms of EDPs such as the roof drift and inter-story drift. The procedure could also be implemented in other SC-MED systems.

REFERENCES

Chou, C.C., Chen Y.C., Pham D.H., and Truong V.M. (2014), "Steel braced frames with dual-core SCBs and sandwiched BRBs: Mechanics, modeling and seismic demands," *Engineering Structures*, 72, 26–40.

Fajfar P. (2000), "A nonlinear analysis method for performance-based seismic design," *Earthquake Spectra*, 16(3), 573–592.

Liu L. (2012), "Seismic behaviour and design of structure with self-centering buckling-restrained braces," Ph.D. thesis. Harbin, Harbin Institute of Technology. (in Chinese)

Liu L., Zhao J., and Li S. (2018a), "Nonlinear displacement ratio for seismic design of self-centering buckling-restrained braced steel frame considering trilinear hysteresis behaviour," *Engineering Structures*, 158, 199–222.

Liu L., Li S., and Zhao J. (2018b), "A novel non-iterative direct displacement-based seismic design procedure for self-centering buckling-restrained braced frame structures," *Bulletin of Earthquake Engineering*, 16(11), 5591–5619.

Liu L., Liu Y., and Zhu X.J. (2019), "Evaluation of nonlinear displacement of self-centering structures with metallic energy dissipaters by considering the early yield of the energy dissipating system," *Soil Dynamics and Earthquake Engineering*, 125. https://doi.org/10.1016/j.soildyn.2019.105757

Lu X., Wu H., and Zhou Y. (2017), "Seismic collapse assessment of self-centering hybrid precast walls and conventional reinforced concrete walls," *Structural Concrete*, 18(6), 938–949.

Martin A., Deierlein G.G., and Ma X. (2019), "Capacity design procedure for rocking braced frames using modified modal superposition method," *Journal of Structural Engineering*, 145(6), pp 1–16.

Miranda E., and Bertero V.V. (1994), "Evaluation of strength reduction factors for earthquake-resistant design," *Earthquake Spectra*, 10(2), 357–379.

Perez F.J., Sause R., and Pessiki S. (2007), "Analytical and experimental lateral load behavior of unbonded posttensioned precast concrete walls," *Journal of Structural Engineering*, 133(11), 1531–1540.

Priestley M.J.N. (1993), "Myths and fallacies in earthquake engineering-conflicts between design and reality," *Bulletin of the New Zealand National Society for Earthquake Engineering*, 26(3), 329–341.

Qiu C.X., and Zhu S. (2016), "High-mode effects on seismic performance of multi-story self-centering braced steel frames," *Journal of Constructional Steel Research*, 119, 133–143.

Qiu C.X., and Zhu S. (2017), "Performance-based seismic design of self-centering steel frames with SMA-based braces," *Engineering Structures*, 130, 67–82.

Ricles J.M., Sause R., Garlock M.M., and Zhao C. (2001), "Posttensioned seismic-resistant connections for steel frames," *Journal of Structural Engineering*, 127(2), 113–121.

Roldán R., Sullivan T.J., and Corte G.D. (2016), "Displacement-based design of steel moment resisting frames with partially-restrained beam-to-column joints," *Bulletin of Earthquake Engineering*, 14(4), 1017–1046.

Steele T.C., and Wiebe L.D.A. (2016), "Dynamic and equivalent static procedures for capacity design of controlled rocking steel braced frames," *Earthquake Engineering & Structural Dynamics*, 45(14), 2349–2369.

Suarez V.A., and Kowalsky M.J. (2010), "Direct displacement-based design as an alternative method for seismic design of bridges," *Special Publication*, 271, 63–78.

Tena-Colunga A., Hernández-Ramírez H., and de Jesús Nangullasmú-Hernández H. (2019), "Resilient design of buildings with hysteretic energy dissipation devices as seismic fuses," In *Resilient Structures and Infrastructure* (Eds: Norrozinejad, E., Takewaki, I., Astaneh-Asl, A., Gardoni, P.), 77–103, Springer, Singapore

Vidic T., Fajfar P., and Fischinger M. (2010), "Consistent inelastic design spectra: Strength and displacement," *Earthquake Engineering & Structural Dynamics*, 23(5), 507–521.

Wiebe L., and Christopoulos C. (2014), "Performance-based seismic design of controlled rocking steel braced frames. I: Methodological framework and design of base rocking joint," *Journal of Structural Engineering*, 141(9), pp 1–11).

Xue Q., and Chen C.C. (2003), "Performance-based seismic design of structures: A direct displacement-based approach," *Engineering Structures*, 25(14), 1803–1813.

Zhou Z., Xie Q., Lei X.C., He X.T., and Meng S.P. (2015), "Experimental investigation of the hysteretic performance of dual-tube self-centering buckling-restrained braces with composite tendons," *Journal of Composites for Construction*, 19(6), 04015011.

14 A Decision-Making Framework for Enhancing Resilience of Road Networks in Earthquake Regions

Anastasios Sextos and Ioannis Kilanitis

CONTENTS

14.1 Introduction 213
14.2 Overview of the Probabilistic Resilience Framework 215
14.3 Key Components of the Network System 215
14.4 Pre-Earthquake Traffic Conditions 215
    14.4.1 Network Portfolio Value, Repair Cost Ratio, and Traffic Capacity Evolution Relationship 217
    14.4.2 Seismic Hazard Analysis 217
    14.4.3 Fragility Analysis 217
    14.4.4 Traffic Analysis 219
14.5 Seismic Risk Assessment of the "As-Built" Network 219
14.6 Risk Management and Mitigation Strategies 220
14.7 Conclusions 221
References 221

14.1 INTRODUCTION

Intercity transportation networks constitute a vital component of prosperity in modern, densely populated societies by facilitating the mobility of people, goods, and services. Their smooth and undisrupted operation is crucial for ensuring sustainability after extreme natural disasters. Earthquakes, for instance, have caused extensive damage worldwide primarily to seismically sub-standard road or highway components (i.e., bridges, overpasses, tunnels, etc.) [1, 2], as well as to adjacent geotechnical works that influence road functionality (e.g., slopes). These damages have led to enormous direct and indirect loss to the affected areas [3]. Direct loss is related to the repair of the damaged components, if one for the sake of quantification neglects the priceless loss of human life, while indirect loss refers to the reduced functionality of the road network and the subsequent increase of travel time, the disturbance to social and professional life, business interruption, additional transportation cost, and environmental implications [4, 5].

Direct and indirect loss associated with future seismic events affecting highway networks and their secondary roadways is assessed probabilistically [6–9], by coupling structural/geotechnical vulnerability [10–12] with the hazard at the site(s) of interest [13–15], as well as the altered traffic flow [16] and the wider economic, social, and environmental consequences of both infrastructure failure and traffic diversion. Key to assessing the community loss is the concept of network resilience that encompasses the dimensions of network capacity, redundancy, and recovery time to express its ability to withstand and adapt to a natural disaster, while being able to recover and restore the services offered quickly. Following the basic ideas of Bruneau et al. [17], different studies aim to quantify resilience. Most of these efforts consist of models of general use [18, 19] that may be applied to different kinds of complex systems such as hospitals [20], lifelines [21], energy systems [22], and infrastructure networks [23]. Due to the interdependency and complexity of roadway networks, as well as their extension in large geographical areas, quantification of resilience and informed decision-making based on the resilience metrics, face several challenges that hinder the practical application of such innovative concepts in existing roadway networks.

This chapter presents a summary of a robust, multi-dimensional, easily applicable, decision-making framework for the quantification and efficient management of road network resilience to seismic hazard that treats all the prevailing uncertainties [24] with a balanced degree of sophistication. More precisely, it further aims to:

- Develop a semi-probabilistic, multi-event, seismic hazard approach, specifically tailored to the salient features of network resilience analysis.
- Integrate multi-phase traffic analysis, considering that immediately after the natural disaster a critical network component (bridge, tunnel, etc.) may either retain 100% of its traffic carrying capacity, operate partially, or be completely closed. In the latter two cases, the network recovers gradually while traffic adapts to the new network conditions of each post-earthquake phase.
- Introduce resilience-based and cumulative roadway resilience indicators to quantify the degree of satisfaction of pre-defined resilience objectives such as network redundancy, integrity of the network components, access to critical services, environmental aspects, and minimum disruption of the financial activity and commerce.

With the aid of an open GIS-based software ad-hoc developed for this purpose, the above methodological framework permits roadway stakeholders to form alternative risk management strategies to minimize earthquake-induced loss prior to a major seismic event and optimize the time and resources needed to restore functionality after the earthquake disaster.

14.2 OVERVIEW OF THE PROBABILISTIC RESILIENCE FRAMEWORK

The methodology presented herein has been developed in the framework of the research program Retis-Risk (www.retis-risk.eu) for the seismic risk assessment and resilience enhancement of interurban roadway networks. It is programmed as a standalone GIS software and is applied to the case of the interurban roadway network of the Western Macedonia prefecture, in Greece. Herein, the methodology, illustrated in Figure 14.1, is demonstrated directly through the case study to demonstrate both the concept and its applicability. Egnatia Highway A2, is a recently constructed highway that extends from the western port of Igoumenitsa to the eastern Greek–Turkish border running a total of 670 kilometers within a challenging earthquake-prone region with design PGA varying from 0.16–0.24 g. Egnatia Highway crosses the prefecture studied and consists of the backbone of its road network, being complemented by several secondary roads that serve the regional transportation needs (Figure 14.1, left). For the purposes of this pilot application, both the main highway of the region under study and the secondary road system with speed limits lower than 90 kilometers/hour are modeled with a total number of 263 bi-directional links and 283 traffic nodes [25].

14.3 KEY COMPONENTS OF THE NETWORK SYSTEM

The set of key network components, that is, the structures and geo-structures whose failure may lead to road closure, is the first to be identified. Key components are assumed to be the bridges, overpasses, slopes, and tunnels across the network. Given the structure of the interurban system studied, it is only bridges and tunnels that are studied in the particular case study; however, the methodology permits all the classes mentioned above. Overpasses of the secondary network are also neglected for simplicity given their smaller size, simpler structural systems, and minor effect to the overall network resilience. Again, if deemed appropriate, their vulnerability can also be accounted for, both by the methodology and the software developed, which are structure-, size-, and importance-independent. Effectively, any class of highway critical components with a known common probability of exceeding predefined damage states and lead to road closure can be modeled as a distinct class. The same applies to special structures (i.e., long bridges or important tunnels) where a single structure can consist of a class on its own. In the framework of the case studied, a total number of 74 dual branch (i.e., directions) key components were identified within the network system. Since the traffic along each network link is bi-directional, each identified key component comprises of two identical branches with a unique ID number per pair.

14.4 PRE-EARTHQUAKE TRAFFIC CONDITIONS

An Origin-Destination (OD) matrix is used to describe the initial, pre-earthquake, travel demands within the entire network for all possible combinations, extracted from a relevant study carried out by the stakeholder. Given the travel demands and

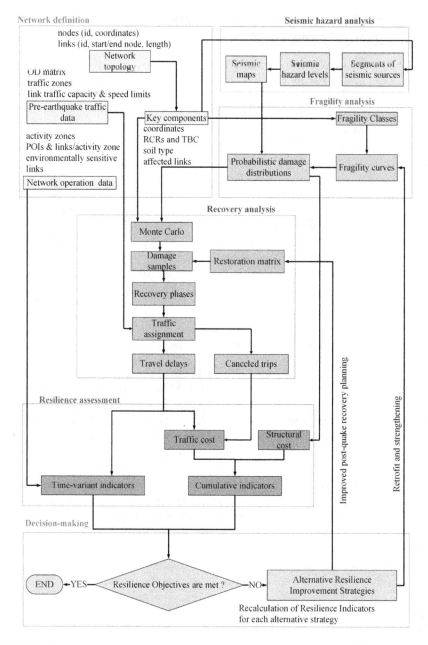

FIGURE 14.1 General workflow of the proposed framework [26].

the additional input of the traffic capacity of every network link, pre-earthquake traffic flows over the whole network are calculated according to Zhou et al. [16] It is noted that the OD matrix used herein, refers to travel demands during the typical hour of a normal day and thus appropriate scaling factors are applied to the results whenever daily traffic data are deemed.

14.4.1 Network Portfolio Value, Repair Cost Ratio, and Traffic Capacity Evolution Relationship

A re-construction cost was calculated for each one of the 74 dual branch key components assuming a value of 17,000€/meter for the (twin) bridges and overpasses and 20,000€/meter for the tunnels. Based on the length of each component, the total value of the network portfolio was approximately assessed to be 630€ million. Moreover, a damage state-specific repair cost ratio was defined for all the key components, according to Basoz et al. [27], assuming ratios of 0.03, 0.25, 0.75, and 1 for Damage State 1 (DS1) to Damage State 4 (DS4), respectively. A closure period of 0, 7, 150, and 450 days is assigned to the four damage states, DS1 to DS4, for all key network components, assuming that after this period, 100% of the traffic carrying capacity is regained.

14.4.2 Seismic Hazard Analysis

The integration of seismicity from different earthquake sources that is expressed in the form of conventional seismic hazard maps, is not applicable for the case of the post-earthquake traffic distribution, as the latter depends on the individual probability of operation of each network key component, which depends on the specific seismic scenario examined and the corresponding spatial distribution of the Intensity Measures (IM) of interest [28, 29]. For this reason, a hazard is herein assessed independently for each one of the m seismic sources potentially affecting the network and for a set of k different return periods defined by the stakeholder. In the case studied, 11 seismic sources ($m = 11$) were identified, located either within the case study area or in its immediate vicinity. For every fault, ground motion maps associated with the $k = 4$ return periods, namely 100, 475, 980, and 1890 years, were generated leading to a sample of $k \times m$ maps depicting the spatial distribution of intensity for every return period and source combination.

14.4.3 Fragility Analysis

For every bridge and overpass key component of this study, a set of four fragility curves was generated for the four damage states considered, namely DS1 to DS4, corresponding to minor, moderate, extensive damage, and collapse, respectively (Figure 14.2, left). Bridges and overpasses are organized in classes of identical fragility, while for important bridges of the network a bridge-specific methodology is followed involving nonlinear static and incremental dynamic response history analysis. For simplicity, the stock of the 28 twin tunnels of the network was grouped into one gross tunnel fragility class based on relationships expressed in terms of peak ground velocity. To be consistent with the PGA-based maps developed, a transformation of PGV to PGA was performed, according to Wald et al. [30]. Figure 14.2 (right) illustrates a sample fragility map showing the most probable damage states of each key component on the basis of the PGA spatial distribution calculated for a sample seismic source and a return period of 475 years. As anticipated for a newly constructed highway that conforms to Eurocode 8, most critical components do not exceed DS1 and DS2 for the particular return period.

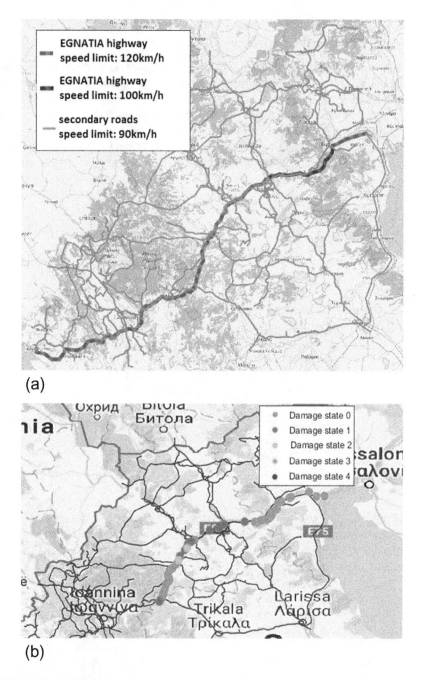

FIGURE 14.2 Sample fragility distribution map showing the most probable DS for every key component (left). Closed links for the recovery phase of the first seven days, for the 475-year map of the "*k*" seismic scenario (right).

14.4.4 TRAFFIC ANALYSIS

Having generated $m = 11$ different seismic maps for each return period, a corresponding set of traffic scenarios is then developed, under the simplifying assumption that immediately after an earthquake a key network component may either retain 100% of its traffic carrying capacity (i.e., remain intact and hence, fully operational) or close and completely lose its traffic capacity. Along these lines, each one of the 74 key components is assumed to perform in a binary manner, associated to a value of either 1 (fully functional) or 0 (closed) based on whether the damage induced exceeds a critical level of damage (in our case $DS_{cr} = DS2$). Given the individual damage state probabilities computed for the critical components per seismic map and return period, a Monte Carlo (MC) analysis is employed assuming ten initial traffic scenario samples that correspond to the state of open and closed network links of each map. Hence, a group of $11 \times 10 = 110$ initial traffic scenarios is generated for each one of the $k = 4$ earthquake return periods. Every initial traffic scenario is then decomposed to several phases that evolve in time based on the stepwise opening of the key components throughout the recovery period.

14.5 SEISMIC RISK ASSESSMENT OF THE "AS-BUILT" NETWORK

The total cost associated with each earthquake event k (k taking values from 1 to 4 for the 100, 475, 980, and 1890 years return period), is the sum of the cumulative direct cost of structural damage within the network and the indirect, earthquake-induced total traffic cost. Based on the repair cost ratios defined and the probability of attaining every damage state, the Estimated Structural Cost $ESC_{k,m}$ due to earthquake with return period k stemming from a source m is derived for each one of the $i = \{1..74\}$ key network components as:

$$ESC_{k,m} = \sum_{i=1}^{74} D_{i,k,m} \quad (14.1)$$

where: $D_{i,k,m} = TBC_i \cdot \left(RCR_1 \cdot P_{DS1}^{i,k,m} + RCR_2 \cdot P_{DS2}^{i,k,m} + RCR_3 \cdot P_{DS3}^{i,k,m} + RCR_4 \cdot P_{DS4}^{i,k,m} \right)$

and TBC_i is the total cost of re-constructing key component i calculated based on its length and the re-construction cost per meter values defined, $\{RCR_1^i, RCR_2^i, RCR_3^i, RCR_4^i\} = \{0.03, 0.25, 0.75, 1\}$ {0.03, 0.25, 0.75, 1} are the repair cost ratios that correspond to Damage States DS1 to DS4, and $P_{DS}^{i,k,m}$ is the probability that the damage of the key component i exceeds DS1 to DS4 for the case of seismic source m and an event return period k. The earthquake-induced *traffic* cost (TC) is then calculated for every Monte Carlo simulated traffic scenario. This cost refers to the *additional* traffic cost during the entire recovery period of that particular traffic scenario (again for each seismic source m and an event return period k), and as such, it is the sum of the product of each phase duration, times the corresponding additional travel cost:

$$TC_{k,m,n_{samp}} = \sum_{p=1}^{P_{k,m,n_{samp}}} EC_{k,m,n_{samp},p} \cdot t_{k,m,n_{samp},p} \quad (14.2)$$

where: $EC_{k,m,n_{samp},p}$: is the *additional* travel cost due to travel delays during phase p of the n_{samp} traffic scenario sampled from the mth IM distribution of earthquake k calculated according to [31], $t_{k,m,n_{samp},p}$: is the duration of phase p of the n_{samp} traffic scenario sampled from the mth IM distribution of event k, $P_{k,m,n_{samp}}$: is the total number of recovery phases associated with n_{samp} traffic scenario sampled from the mth IM distribution of event k. Subsequently, the estimated traffic cost (ETC) can be associated to every seismic map, as the mean of the costs calculated for the ten Monte Carlo samples (i.e., each one for each phase) simulated from that map:

$$ETC_{k,m} = \frac{\sum_{n_{samp}=1}^{10} TC_{k,m,n_{samp}}}{10} \qquad (14.3)$$

The maximum estimated structural and traffic cost out of the 11 cases of individual seismic sources leads to the envelope total network cost (TNC_k) and identifies the *critical seismic source* that has a higher contribution to the overall loss.

14.6 RISK MANAGEMENT AND MITIGATION STRATEGIES

Having identified the direct and indirect cost associated with every return period as well as the qualitative resilience indicators reflecting the wider consequences of the earthquake event, alternative risk mitigation strategies are developed and comparatively assessed. A retrofit scheme is developed for the particular bridges leading to updated fragilities or reduced probability of failure for the same Intensity Measure. The updated fragilities were, in this case, approximately derived by multiplying the mean threshold value of the corresponding "as-built" components by 1.3 for all DSs. A second risk management strategy consisting of improved post-earthquake response expressed through an improved traffic carrying capacity-time relationship was also considered. In this case, closure periods were assumed to be lower due to better recovery planning and were updated to 0, 4, 100, and 300 days instead of 0, 7, 150, and 450 days for Damage States 1 to 4, respectively. Figure 14.3 depicts the resulting estimated structural, traffic, and total cost for different earthquake return periods for the case of the "as-built" network as well as the two risk management strategies (i.e., bridge retrofit or improved recovery planning) due to the seismic maps derived from the critical seismic source. Retrofit of selected key components is found more effective compared to the recovery plan enhancement for all the examined return periods. This is because, in this particular network, structural cost, which is essentially unaffected by an improved recovery, is much higher than traffic cost. However, both risk management strategies contribute to a non-negligible, yet small (5–18%), extent to the estimated total network cost reduction again due to the high resilience and low expected loss of the "as-built" network.

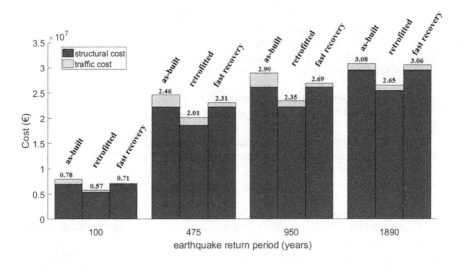

FIGURE 14.3 Expected costs for the four seismic scenarios for the case of the "as-built" network and the two risk mitigation strategies.

14.7 CONCLUSIONS

A holistic framework is presented for the multi-criterion assessment and management of the seismic resilience of roadway networks through a balanced treatment of different sources of uncertainty that contribute to the overall network seismic risk, namely, vulnerability and hazard, coupled with refined traffic and consequences analysis. Several novel, resilience-based, and cumulative indicators are introduced for reflecting the time-variant nature of losses and providing a compact and more easily to interpret estimate of the extent of the losses incurred by the community after a major seismic event as well as throughout the entire recovery period. The holistic estimate of the direct (structural) and indirect (traffic) monetary loss as well the wider financial, network connectivity, and environmental impacts of seismic events with different probability of occurrence are used as a means to describe the evolution of network functionality and resilience taking explicitly into account its gradual recovery. The proposed framework is implemented into a GIS-based software and constitutes a useful decision-making tool for the stakeholders, to quantify and improve the resilience of their roadway network, by testing and adopting the most appropriate alternative resilience improvement strategy.

REFERENCES

1. Kawashima K, Buckle I. Structural performance of bridges in the Tohoku-oki earthquake. *Earthquake Spectra* 2013; **29**(S1): S315–S338.
2. Shen Y, Gao B, Yang X, Tao S. Seismic damage mechanism and dynamic deformation characteristic analysis of mountain tunnel after Wenchuan earthquake. *Engineering Geology* 2014; **180**: 85–98. DOI: 10.1016/j.enggeo.2014.07.017.
3. Zhou Y, Banerjee S, Shinozuka M. Socio-economic effect of seismic retrofit of bridges for highway transportation networks: A pilot study. *Structure and Infrastructure Engineering* 2010; **6**(1–2): 145–157. DOI: 10.1080/15732470802663862.

4. Kiremidjian AS, Moore J, Fan YY, Yazlali O, Basoz N, Williams M. Seismic risk assessment of transportation network systems. *Journal of Earthquake Engineering* 2007; **11**(3): 371–382. DOI: 10.1080/13632460701285277.
5. Enke D, Tirasirichai C, Luna R. Estimation of earthquake loss due to bridge damage in the St. Louis metropolitan area. II: Indirect losses. *Natural Hazards Review* 2008; **9**(1): 12–19.
6. Chang S, Shinozuka M, Moore JE. Probabilistic earthquake scenarios: Extending risk analysis methodologies to spatially distributed systems. *Earthquake Spectra* 2000; **16**(3): 557–572.
7. Dong Y, Frangopol DM, Saydam D. Sustainability of highway bridge networks under seismic hazard. *Journal of Earthquake Engineering* 2014; **18**(1): 41–66. DOI: 10.1080/13632469.2013.841600.
8. Werner SD. A risk-based methodology for assessing the seismic performance of highway systems. *MCEER* 2000: 63–71.
9. Günneç D, Salman FS. Assessing the reliability and the expected performance of a network under disaster risk. *OR Spectrum* 2011; **33**(3): 499–523. DOI: 10.1007/s00291-011-0250-7.
10. Kim SHH, Feng MQMQ. Fragility analysis of bridges under ground motion with spatial variation. *International Journal of Non-Linear Mechanics* 2003; **38**(5): 705–721. DOI: 10.1016/S0020-7462(01)00128-7.
11. Desroches R, Padgett JE, Nilsson E. Retrofitting transportation systems to ensure resiliency. *14th World Conference of Earthquake Engineering* 2008.
12. Shafieezadeh A, DesRoches R, Rix GJ, Werner SD. A probabilistic framework for correlated seismic downtime and repair cost estimation of geo-structures. *Earthquake Engineering & Structural Dynamics* 2014; **43**(5): 739–757. DOI: 10.1002/eqe.
13. Han Y, Davidson RA. Probabilistic seismic hazard analysis for spatially distributed infrastructure. *Earthquake Engineering & Structural Dynamics* 2012; **41**(15): 2141–2158. DOI: 10.1002/eqe.
14. Bommer JJ, Crowley H. The influence of ground-motion variability in earthquake loss modelling. *Bulletin of Earthquake Engineering* 2006; **4**(3): 231–248. DOI: 10.1007/s10518-006-9008-z.
15. Sokolov V, Wenzel F. Influence of ground-motion correlation on probabilistic assessments of seismic hazard and loss: Sensitivity analysis. *Bulletin of Earthquake Engineering* 2011; **9**(5): 1339–1360. DOI: 10.1007/s10518-011-9264-4.
16. Zhou X, Taylor J, Pratico F. DTALite: A queue-based mesoscopic traffic simulator for fast model evaluation and calibration. *Cogent Engineering* 2014; **1**(1): 961345. DOI: 10.1080/23311916.2014.961345.
17. Bruneau M, Chang SE, Eguchi RT, Lee GC, O'Rourke TD, Reinhorn AM, et al. A framework to quantitatively assess and enhance the seismic resilience of communities. *Earthquake Spectra* 2003; **19**(4): 733–752. DOI: 10.1193/1.1623497.
18. Zobel CW, Khansa L. Characterizing multi-event disaster resilience. *Computers & Operations Research* 2014; **42**: 83–94. DOI: 10.1016/j.cor.2011.09.024.
19. Cimellaro GP, Renschler C, Reinhorn AM, Arendt L. PEOPLES: A framework for evaluating resilience. *Journal of Structural Engineering* 2016; **142**(10). DOI: 10.1061/(ASCE)ST.1943-541X.0001514.
20. Cimellaro GP, Reinhorn AM, Bruneau M. Seismic resilience of a hospital system. *Structure and Infrastructure Engineering* 2010; **6**(1–2): 127–144. DOI: 10.1080/15732470802663847.
21. Cimellaro GP, Villa O, Bruneau M. Resilience-based design of natural gas distribution networks. *Journal of Infrastructure Systems* 2014: 1–14. DOI: 10.1061/(ASCE)IS.1943-555X.0000204.

22. Didier M, Grauvogl B, Steentoft A, Ghosh S, Stojadinović B. Seismic resilience of the Nepalese power supply system during the 2015 Gorkha earthquake. *16th World Conference on Earthquake Engineering*, Santiago, Chile, January 9–13, 2017.
23. Ouyang M, Dueñas-Osorio L, Min X. A three-stage resilience analysis framework for urban infrastructure systems. *Structural Safety* 2012; 36–37: 23–31. DOI: 10.1016/j.strusafe.2011.12.004.
24. Biondini F, Frangopol DM. Life-cycle performance of structural systems under uncertainty. *ASCE Journal of Structural Engineering* 2016; **142**(9): 2233–2244. DOI: 10.1061/9780784413357.196.
25. Sextos AG, Kilanitis I, Kappos AJ, Pitsiava M, Sergiadis G, Margaris V, et al. Seismic resilience assessment of the western Macedonia highway network in Greece. *6th ECCOMAS Thematic Conference on Computational Methods in Structural Dynamics and Earthquake Engineering M. Papadrakakis, M. Fragiadakis (eds.)*, Rhodes Island, Greece, June 15–17, 2017.
26. Kilanitis I, Sextos AG. Integrated seismic risk and resilience assessment of roadway networks in earthquake prone areas. *Bulletin of Earthquake Engineering*, 2019; **17**(1): 181–210.
27. Basoz N, Kiremidjian AS, King SA, Law KH. Statistical analysis of bridge damage data from the 1994 Northridge, CA, earthquake. *Earthquake Spectra* 1999; **15**(1): 25–54.
28. Bommer JJ, Crowley H. The influence of ground-motion variability in earthquake loss modelling. *Bulletin of Earthquake Engineering* 2006; **4**(3): 231–248. DOI: 10.1007/s10518-006-9008-z.
29. Sokolov V, Wenzel F, Jean WY, Wen KL. Uncertainty and spatial correlation of earthquake ground motion in Taiwan. *Terrestrial, Atmospheric and Oceanic Sciences* 2010; **21**(6): 905–921. DOI: 10.3319/TAO.2010.05.03.01(T).
30. Wald DJ, Quitoriano V, Heaton TH, Kanamori H. Relationships between peak ground acceleration, peak ground velocity, and modified mercalli intensity in California. *Earthquake Spectra* 1999. DOI: 10.1193/1.1586058.
31. Sextos AG, Kilanitis I, Kyriakou K, Kappos AJ. Resilience of road networks to earthquakes. *16th World Conference on Earthquake Engineering*, Santiago, Chile, January 9–13, 2017.

Index

A

AEM based software, 179
Anchorage earthquake (2018 Mw 7.0), 157
ANN. *see* Artificial Neural Network
ANSYS finite element (FE) software, 180
Applied element method (AEM), 179, 180
Applied Technology Council (ATC), 89
Artificial intelligence (AI), 5, 6, 17, 21, 22, 112
Artificial Neural Network (ANN), 112, 116
"As-built" network, 219–221
ATC. *see* Applied Technology Council

B

BCP. *see* Business Continuity Plan
Big data (BD), 7, 31–32, 116
   applications, 11
   characteristics, 11, 12
   machine learning and AI
      challenges, 35
      concepts, 38–39
      description, 31–32
      frameworks, 37–38
      methods, 38–39
      motivation, 32–34
      objectives, 34–35
      overview, 32–34
      roadmaps and strategies, 39
      state of the art, 36–37
      technologies, 38–39
Bridges, resilience of
   AEM, 179, 180
   benefits
      structural control system, 183–184
      structural health monitoring, 184–186
   cable-stayed-bridge, 178
   guidelines, 178
   model, 180, 181
   NTSB, 179
   remarks, 182–183
   robustness and redundancy issues, 181–182
   seismic response simulation, 182
   SHM, 179
   structural behavior, 178
   structural collapses, 178
   structure, 180
Building Seismic Safety Council (BSSC), 89
Business Continuity Plan (BCP), 163

C

Central Business District (CBD), 46, 50
Christchurch
   CBD, 49
   concrete structures, 55
   earthquake, 49
   objective, 55
   qualitative findings and resilience considerations
      decision-making process, 54
      engineering expertise and capabilities, 55
   quantitative findings
      lateral-load resisting systems, 50, 53–55
      steel structural systems, 50–53
CIF. *see* Construction Innovation Forum
CIM. *see* City Information Model
CityGML, 17
City Information Model (CIM), 153
Concrete structures. *see* Fiber-reinforced polymers
Construction Innovation Forum (CIF), 58
Conventional method, 149–151
Coulomb Friction Principle, 129
CS. *see* Concrete structures

D

D'Alembert's Principle, 129
Damage probability matrix (DPM) method, 142
Data mining (DM), 29, 112, 116
DBD. *see* Displacement-based design
DDBD. *see* Direct displacement-based design
Direct displacement-based design (DDBD), 205
Displacement-based design (DBD), 204, 205, 207
DM. *see* Data mining
DPM. *see* Damage probability matrix

E

Early warning systems (EWSs), 26
Earthquake
   emergency response, 142–159
   road networks. *see* Road networks
Eco-system, 83
EDP. *see* Engineering demand parameters
EI. *see* Energy infrastructure
Emergency Services (ES), 22
Empirical models, 142
Energy infrastructure (EI), 22

225

Engineering demand parameters (EDP), 149, 205, 208
Engineering systems seismic resilience
　definition, 163
　emergency functionality analysis
　　definition, 168
　　fault tree, 169
　　idealized repair path, 169, 170
　　linear recovery model, 173
　　Monte-Carlo simulation, 171
　　recovery time, 171
　　residual functionality, 171, 172
　framework, 164–165
　model, 164
　system model and assessment
　　challenges, 166
　　Monte-Carlo simulation, 166, 168
　　state tree method, 166, 167
ES. *see* Emergency Services
EWSs. *see* Early warning systems

F

Fault tree model, 167, 169, 174
Federal Emergency Management Agency (FEMA), 89, 163
　building performance levels, 91
FEMA P-58 method
　building information models, 152
　challenges, 151
　CIM, 153
　field survey data and building design drawings, 152
　GIS database, 152
　process, 151
Fiber-reinforced polymers (FRPs)
　bridge repair, 193–194
　deficient bridge structure, 190–191
　deficient industrial structure, 198–200
　first-generation corrosion free bridge decks, 57–58
　mechanical and physical characteristics, 92
　post-repair work
　　experiments, 194–196
　　test data, 196–198
　rehabilitation techniques, development of, 191–193
FPR-steel reinforced structures
　bond-based design, 101–104
　innovative resilient systems, 104–105
　longitudinal and steel reinforcement, 99–101
　SFCBs, 99
Fragility, 115
FRPs. *see* Fiber-reinforced polymers
Future infrastructure risk control and resilience
　challenges, 13–14
　concepts, methods, and technologies, 16–17
　description, 12
　frameworks, 15–16
　objectives, 12–13
　research gaps, 14–15
　roadmaps and strategies, 17
　solutions, 18

G

General Services Administration (GSA) guidelines, 178
GFRP. *see* Glass fiber reinforced polymer
GIS-based software ad-hoc, 214, 215, 221
Glass fiber reinforced polymer (GFRP), 192, 194, 196–198
　field demonstration projects, 63–65
Ground motion prediction equations (GMPE), 142
GSA. *see* General Services Administration

H

Healthcare–bridge network system, 45
High-performance computing (HPC), 153
Hospital systems, 168; *see also* Emergency functionality analysis
HPC. *see* High-performance computing

I

Immediate Resilience, 183–184
Information technologies (IT), 22
Infrastructure 3.0, 109
Innovation, 83
Input-output inoperability, 116–118
Input-output model, 116
Intelligence, 4
Intercity transportation networks, 213
International Organization for Standardization (ISO), 86–87
International workshop on resilience (IRW), 2, 11
ISIS Winnipeg Principles
　application
　　first-generation corrosion free bridge decks, 57–59
　　Salmon River Highway Bridge, Nova Scotia, 58–60
　second-generation corrosion-free bridge decks, 62
　　fatigue testing, 63
　　GFRP, field demonstration projects, 63–65
　　performance, 63
　second-generation steel-free bridge decks, 60
　　Red River Bridge, 60–62

Index

ISO. *see* International Organization for Standardization
Isolation-structure systems
  description, 123
  FPB model, 124
  seismic performance
    prefabricated structure system, 135–138
    SLDFPB (*see* Super-large displacement friction pendulum bearing)
IT. *see* Information technologies

J

Japan Society of Civil Engineering (JSCE), 89
Jiuzhaigou earthquake (2017 M7.0), 155–157
JSCE. *see* Japan Society of Civil Engineering

K

Kintaikyo Bridge, 68
  history, 67
  objectives, 77
  resilience structures, 69–70
    damage reduction countermeasures, 74–77
    durability countermeasures, 74
    safety countermeasures, 70–72
    serviceability countermeasures, 72–74
    steel bands, 67–68
    voussoir arch method, 68

L

Lagrange Principle, 125, 129
Life-cycle cost (LCC), 82, 84
Life-cycle resilience
  description, 43–44
  functionality, 44–45
  PEB, 46
  quantification, 44–46
  representation, 44
  sustainability, 46

M

MAC. *see* Modal Assurance Criterion
Machine learning (ML), 5–7, 31–33, 35, 37, 112
MCEER. *see* Multidisciplinary Center for Extreme Event Research
MDOF. *see* Multiple-degree-of-freedom
Modal Assurance Criterion (MAC), 180, 181
Multidisciplinary Center for Extreme Event Research (MCEER), 15, 115
Multiple-degree-of-freedom (MDOF), 145, 147, 151, 159
  flexural-shear model, 145
  shear model, 145, 147

N

National Earthquake Hazards Reduction Program (NEHRP), 86, 89, 90
National Research Council (NRC), 163
National Transportation Safety Board (NTSB), 179
Near-real-time earthquake loss estimation tools, 142
NEHRP. *see* National Earthquake Hazards Reduction Program
Network model, 113–114
NRC. *see* National Research Council
NTSB. *see* National Transportation Safety Board

O

Open System for Earthquake Engineering Simulation (Open SEES) software, 98
Origin-Destination (OD) matrix, 215, 216
output-only techniques, 185–186

P

Parameter determination method
  backbone curve based on HAZUS database
    ground motions, 149
    human sense, 149
    MDOF shear model, 147
    pinching model, 148
    post-earthquake emergency response, 149
  in China
    description, 145
    engineering designed structures, 146
    non-engineered buildings, 147
PBE. *see* Performance-based engineering
PBSD. *see* Performance-based seismic design
Performance-based engineering (PBE), 46
Performance-based seismic design (PBSD), 204
Posttensioning Institute (PTI), 178
PPD-8. *see* Presidential Policy Directive 8
Prefabricated structure system, 135–138
Presidential Policy Directive 8 (PPD-8), 43
PTI. *see* Posttensioning Institute

Q

Quality, 115
Quantification of resilience capabilities, 114–115

R

R-C_R DDBD, 204–205
  SC-MED structure, 205–210
RC structures
  FRP composites application

mechanical and physical
characteristics, 92
retrofitting techniques, 91
FRP composites recoverability, 92–95
bond-based design, 101–104
innovative resilient systems, 104–105
limit states, 96–98
longitudinal and steel reinforcement, 99–101
SFCBs, 99
sustainability performance-based design
current situation, 87
definition, 81
global efforts, 85–87
recoverability *vs.* resilience *vs.* sustainability interrelation, 84–85
structures, 82
upgrade structures importance, 87
upgrade structures recoverability/restorability
NEHRP guidelines, 89, 90
performance, 89, 90, 91
seismic design codes, 88
Real-time city-scale nonlinear time-history analysis
earthquake emergency response, applications
Anchorage earthquake (2018 MW 7.0), 157
Jiuzhaigou earthquake (2017 M7.0), 155–157
RED-ACT system, 155
estimation tools, 142
framework, 143, 144
ground motion records, 143
HPC post-earthquake emergency response, 153
MDOF model, 145
multi-story buildings, 145
parameter determination methods, 145
backbone curve on HAZUS data, MDOF shear model, 147–149
in China, 145–147
regional seismic
conventional method, loss prediction, 149–151
FEMA P-58 method, regional resilience assessment, 151–153
seismic loss estimation systems, 142
tall buildings, 145
Real-time Earthquake Damage Assessment using City scale Time-history analysis (RED-ACT), 142, 153, 155, 157
applications, 155
determination, 157–159
Recoverability/restorability, 82
RED-ACT. *see* Real-time Earthquake Damage Assessment using City scale Time-history analysis

Red River Bridge, 60–62
Redundancy, 44, 69–70, 82, 177
in bridge, 69, 70, 181–182
definition, 177
Reliability, 82
resilience measures, 115–116
Resilience of critical infrastructure system;
see also individual entries
big data, machine learning and AI
challenges, 35
concepts, 38–39
description, 31–32
frameworks, 37–38
methods, 38–39
motivation, 32–34
objectives, 34–35
overview, 32–34
roadmaps and strategies, 39
state of the art, 36–37
technologies, 38–39
challenges, 23–24
definitions and objectives, 21–23
future infrastructure risk control and resilience (*see* Future infrastructure risk control and resilience)
health monitoring and early warning, 25–27
innovation technologies, 28–29
intra and interdependencies, 24–25
modeling and simulation, 25
performance, 43
resilience-based design, 27–28
resilience-based improvements, 27
road-mapping, 6–10
amendments, 7–8
endeavors, 7
generic questions, 8–9
group report, 10
guiding question complexes, 6
roadmaps and strategies, 29–30
socio cyber-technical self-learning infrastructure systems (*see* Resilient socio cyber-technical self-learning infrastructure systems)
Resilience quantification
functionality, 44
hazards, 45, 46
healthcare–bridge network system, 45
performance, 45
recovered state, 45
recovery state, 45
reliability state, 44, 45
Resilient socio cyber-technical self-learning infrastructure systems
acceptance, 4
artificial intelligence, 5
damaging events, 4
feasible combinations, 4

Index

machine learning, 5
properties, capabilities/technical capabilities, 4, 5
resilience cycle phases, 3
technological capabilities, 5
Resourcefulness, 44, 82
Retis-Risk, 215
Road networks
  components, 216
  direct and indirect loss, 214
  framework, 215, 216
  intercity transportation networks, 213
  interdependency and complexity, 214
  pre-earthquake traffic conditions, 215–216
    fragility analysis, 217–218
    portfolio value, 217
    repair cost ratio, 217
    seismic hazard analysis, 217
    traffic analysis, 219
    traffic capacity evolution relationship, 217
  risk management and mitigation strategies, 220, 221
  seismic risk assessment, 219–220
Robustness, 82–83, 114, 177
  in bridge, 181–182
  definition, 177

S

Salmon River Highway Bridge, Nova Scotia, 58–60
SC buckling-restrained braced frame (SC-BRBF)
  elevation, 206
  linear and nonlinear demand, 208
  linear pushover analysis, 207
  nonlinear pushover analysis, 207, 208
SC-MED structure
  concept and restoring force, 204
  description, 203
  hysteresis, 206
  R-C_R DDBD application, 205–210
Self-centering (SC) structure, 203
SFCBs. *see* Steel fiber composite bars
SHM. *see* Structural control and health monitoring
6-systems model, 116–117
SLDFPB. *see* Super-large displacement friction pendulum bearing
Socio cyber-technical self-learning infrastructure systems
  acceptance, 4
  artificial intelligence, 5
  damaging events, 4
  feasible combinations, 4
  machine learning, 5
  properties, capabilities/technical capabilities, 4, 5
  resilience cycle phases, 3
  technological capabilities, 5
State tree method, 166, 167, 174
Steel fiber composite bars (SFCBs), 99
Structural control and health monitoring (SHM), 179
Super-large displacement friction pendulum bearing (SLDFPB)
  isolation structure, 130–135
    parameters, 131
    rare and super-strong earthquakes response, 132
  mechanical properties, 125–130
    algebraic sum, 129
    Coulomb Friction Principle, 129
    D'Alembert's Principle, 129
    frame structure, 128
    isolation structure, 126
    Lagrange Principle, 125
    rare and super-strong earthquakes response, 132
    superstructure, 125
Super-large displacement translation friction pendulum bearing, 138–140

T

3D isolation system, 124
TI. *see* Transportation infrastructure
Timber bridge. *see* Kintaikyo Bridge
Transportation infrastructure (TI), 22

U

Urban Infrastructures Resilience Assessing
  BD, 110–112
  challenges, 112–113
  DM, 112
  methodology
    input-output inoperability, 116–118
    input-output model, 116
    network model, 113
    quantification capabilities, 114–115
    resilience measures, 115–116
  resilience, 110, 111
  restoration analysis, 118, 119
  RR and RRI analysis, 119–120
Urban structures, 80–81

W

Waste water system (WW), 22
World Conference on Disaster Reduction (WCDR), 163

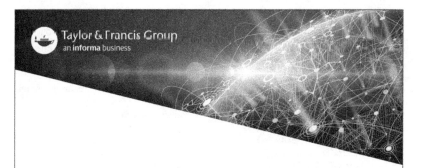

CPSIA information can be obtained
at www.ICGtesting.com
Printed in the USA
BVHW051546060323
659797BV00002B/200